Endophysics

The World as an Interface

Endophysics
The World as an Interface

Otto E. Rössler
Universität Tübingen

Preface: Peter Weibel
Text selection: Peter Weibel
Editing: Reimara Rössler

World Scientific
Singapore • New Jersey • London • Hong Kong

Published by

World Scientific Publishing Co. Pte. Ltd.

5 Toh Tuck Link, Singapore 596224

USA office: 27 Warren Street, Suite 401-402, Hackensack, NJ 07601

UK office: 57 Shelton Street, Covent Garden, London WC2H 9HE

British Library Cataloguing-in-Publication Data
A catalogue record for this book is available from the British Library.

ENDOPHYSICS
The World as an Interface

ISBN-13 978-981-02-2752-4
ISBN-10 981-02-2752-3

Dedicated to Hans Weil

on the

occasion of his 95th birthday

Contents

Preface

In this book a new science, endophysics, is brought to the reader's attention. Following relativity, quantum mechanics and chaos theory, for the fourth time in a century a radical questioning of the understanding of reality takes place. Relativity has suspended the absoluteness of space and yet, at the same time, subjected all phenomena to the velocity of light as an ultimate absolute constant. Quantum mechanics, by introducing the observer, has relativized the objective character of the world. Chaos theory amplified the unavoidable error of measurement and has thereby rendered the unpredictability of the future an inescapable fact. After the "relativization" of objectivity through an absolute speed limit, observation and unpredictability, endophysics brings the chain to its logical end by replacing the traditional assumption of the external observer (exophysics) with the introduction of the internal observer. This internal perspective differs from the quantum-physical problem of an observer dependence in that the measurement problem is in quantum mechanics still believed to be objective (and in reality to be objectively capricious), while in endophysics the internal observer is constitutive in an exact sense. The relativization and observer dependence of the world thereby becomes considerably more radical. Chaos theory, too, still presupposes a (prequantum) external observer. Whereas in chaos theory the unpredictability follows from the error explosion, the loss of attainable knowledge in endophysics is not error-induced but structural.

Endophysics shows us to what extent objective reality is necessarily dependent on the observer. Ever since the introduction of perspective during the Renaissance and of group theory in the 19th century, we have known that the appearances of the world depend in a lawful manner on the localization of the observer ("codistortion"). Only if one is completely outside a complex universe is a complete description of the latter possible (cf. Gödel). According to endophysics, it is only in a model that this position on the outside of a complex universe is possible, but *not* in reality itself. Endophysics hence provides an approach to a general model and simulation theory (and also to the "virtual realities" of the computer age). It is an outgrowth of chaos theory, to which Otto Rössler has contributed since 1975 (cf. the celebrated Rössler attractor, 1976). Another aspect of endophysics lies in the production of novel interpretations of quantum-physical problems. Rössler builds a bridge between the quantum-physical interpretations of Everett, Bell and Deutsch on the one hand and the fractal mechanics of Nelson, Nagasawa, Ord, Prigogine and El Naschie on the other.

Endophysics is different from exophysics, for the physical laws which apply when one is a part of that which one contemplates are in general not the same as those which are valid from an imagined or real external vantage point. Gödel's undecidability also holds true only from within — from the inside of the mathematical system in question, that is.

In physics, one needs to incorporate an explicit observer into the model world, in order to make the reality which exists for him or her understandable. Endophysics entrusts us, as it were, with a "double access" to the world. Besides the direct access (through the interface of the senses), a second one is opened up based on an imaginary outside position. Is the so-called objective reality only the endo side of an exo world?

Time and again, the history of cultural productivity yields evidences that human beings divine the possibility that the world is only the endo side of an exo world. This hypothesis manifests itself in numerous pictorial representations, gnostic formulations, riddles and paradoxes. To illustrate the concept of the interface as the only reality, the model of the "Bubble Boy" is suggestive — he lives in a sterile plastic chamber with floating walls, communicating with the world only through the interface. The menu of his world programming is located on the keyboard inside the bubble. Our own macroscopic world is irreversible, but the bubble in which we reside is microscopically reversible — with counterintuitive consequences.

The fact that our world at the same time is nonclassical is not necessarily an objection. Indeed the classical time reversal invariance and the classical permutation invariance, enjoyed by equal classical particles, lead to "nonclassical" nonlocal phenomena. The "rest of the universe" is for the internal observer distorted in an incorrigible fashion. The world is made of rubber, but we do not realize this because we are made of rubber, too. The implied "simultaneity hypersurfaces" are, from the viewpoint of an external observer, warped in a most complicated fashion. The external observer feels tempted to drop "hints" to the internal observer that would enable him or her to glance behind the curtain. Unfortunately, we do not possess in our own world a similar "giant eye" which we might turn to for help. Unless, that is, we seek refuge in the construction of a fictitious all-knowing, all-powerful super-observer.

The only scientific method for finding out whether or not our own world possesses an exo-objective flip side is the construction of model worlds (or artificial universes) on a level that lies below our own world. This way of proceeding is called endophysics.

The classical picture of an objective world (predictability, explicability, invariance, closure, causality, locality) can, nevertheless, give rise to nonclassical observer-centered effects, as Einstein proved with his special theory of relativity. The new endophysical principle of observer relativity (in place of mere frame relativity) will, perhaps, lead to a theory of Now and Death. Descartes' program of springing the prison of the *hic et nunc* through analytic geometry and the algebraization of experience is updated through the introduction of the observer as a source of covarying distortion. The manipulability of the jail of space and time grows by a notch. The world taken as a "repair shop for the wish machine" takes a step forward.

The endo approach offers a promise to the complex technoworld of the electronic epoch. The effects of the industrial (machine-based) and postindustrial (information-based) civilization — machinization, mediatization, simulation, synthetization, semiosis, artificial reality, deprivation of existence, etc. — are all drawn into a new discourse. The endo approach provides a new theoretical frame for describing and understanding the scientific, technological and social conditions of the postmodern world. For instance, the virtual worlds form a special case of endophysics. The issues raised by endophysics — ranging from observer relativity via the representational paradox and nonlocality to the problem of the world as pure interface — are central questions for our electronic–telematic civilization.

Peter Weibel
Vienna

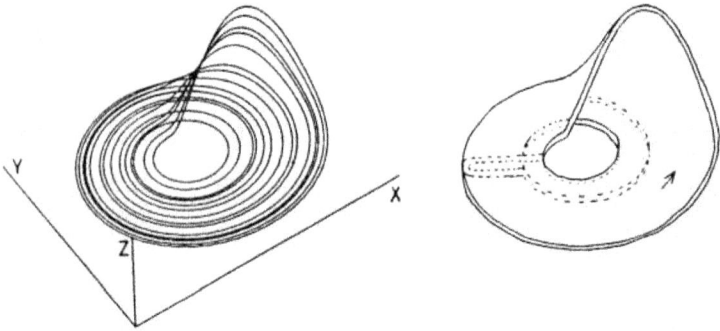

חהו ובהו   ΧΑΟΣ   混 沌

Fig. 0. Chaos.

1
Anaxagoras' Idea of the Infinitely Exact Chaos

Summary

A connection is drawn between ideas of the pre-Socratic philosopher Anaxagoras, on the one hand, and some known or not so well-known facts about chaos in the modern mathematical sense, on the other. In particular, the "long line" and "cloudlike" objects will be mentioned. A new hierarchy as a function of dimensionality is suggested.

1.1 Introduction

Anaxagoras is the most noteworthy, perhaps, for his wish that in the month of his death, schoolchildren would be given a holiday, a custom which was observed for centuries.[1] In second position comes his discovery of dynamical systems (deterministic flows) and of transfinite accuracy. Third, he introduced conservation and time reversal invariance into physics. Fourth, he invented dualism and the idea of a ghost-free world. For the latter view he was persecuted and exiled.

All of this (plus a few more details) is easy to verify from his writings, since only about four pages have survived from the two-volume book on physics (*Peri Physeos*, i.e. "About Nature") which he wrote. One of the fragments — the largest and perhaps most informative one — is reproduced here (see Appendix). The translation is a contemporary one, which may be seen from the fact that the term "self-similar" appears in it.

The reason Anaxagoras is of interest in the context of chaos theory is, if you wish, a didactic one. If our current views on chaos can be recast in terms that are two-and-a-half thousand years old, they possess a measure of universality.

1.2 Mixing in Mythology

To put Anaxagoras in perspective, it is instructive to ask what his predecessors knew about the all-important phenomenon of mixing — that is, about kneading dough, battering butter out of milk and drawing sausage out of another life-giving liquid. In China, Hun-Tun (Kon-Ton in Japanese[2]) was the first Emperor of the Middle Empire, his name meaning "mixture" and "chaos."[3] He lacked sense organs, and when his two royal friends from the outer kingdoms had pityingly operated on him — drilling the necessary seven holes into his head in so many days — he died. Kon-Ton today still signifies that state of mind in which you do not need to open your eyes to check on the details — pure understanding.[2] The names Shu and Hu of the two friends who performed the act of pity mean, when taken as a single word (Shu-Hu), "lightning."[3] When we encounter lightning today, we are still permitted a glimpse of the preceding (Hun-Tun) state of the world if we are lucky enough to peek through the crack.[3] In a third myth Hun-Tun means a leather bag filled with blood — and a fiery arrow must not be sent against it.[3]

In India, there is the analogous picture of the god Vishnu, who lies as a tortoise on the bottom of an ocean of milk, juggling a holy rod-shaped mountain (Mandara) on his back and rotating it. This unmixing process creates the moon and all the living forms (cf. the old painting reproduced in Ref. 4).

In the Middle East is Tohuwabohu, a word which since early Hellenistic times has been translated into "chaos." Immediately after this notion is introduced in the second sentence of the Bible, the hovering of the Mind over the waters creates everything. Unlike what holds true in Greek mythology, the mind here is female (Ruah) and the waters (Mayim) are male since the latter word also means "sperm."

In another equally old story (kindly pointed out to me by George Marx) we have the myth of the brothers Osiris and Set in ancient Egypt, with the former cut to pieces by the latter but with the world emerging from the bloody remnants after they have taken part in an act of impregnation.

Into this circle of mutually related myths, Anaxagoras' own fits neatly. ΧΑΟΣ in Greek mythology was the first primordial entity (according to Hesiodus), right before Gaia and Eros. However, in that early time, the word "chaos" apparently had nothing to do with disorder or mixture but signified only the big yawn, the gaping primordial abyss (related to the word chaskein, i.e. "yawning"). Four centuries later, however, Anaxagoras introduced his own foremost primordial entity, a state of perfect mixture. While we have no proof that he actually called it "ΧΑΟΣ" again, the fact that exactly since his time the word "chaos" has assumed the modern meaning (cf. Ref. 5) constitutes an indirect hint.

As this is the case in the above-mentioned myths, also in Anaxagoras' writing a second primordial entity looms large beside the Chaos (or Great

Mixture): the Mind. The NOYΣ is the only thing that is too "fine" to be miscible.[5] This entity acts like an arrow once more, creating at one point in space and time the beginnings of an unmixing process — out of which emerged everything we have seen to date — according to Anaxagoras (see Appendix).

1.3 The Theory Behind the Myth

Anaxagoras' myth shares with the older ones the male–female dualism — the chalice-and-blade structure (to adopt Riane Eisler's phrase). The difference between Anaxagoras' cosmogony and the earlier ones lies in the fact that the language used by him is that of mathematics and physics. His myth is a scientific one. There is only conservation, i.e. no creation in time. Everything had existed forever in an invariant state. This invariant state had to correspond to that of a perfect mixture. But then the problem arose of how one could explain the emergence of the simple things as we see them today from the complex — the inverse problem to the usual explanatory models. At this point a second principle was needed: that too fine and light substance, the Mind. It initiated a lawful process of unmixing. Unmixing a perfect mixture is, perhaps, the ultimate impossible act. Chaos and ordering are polar opposites. A totally mixed state, obtained after an infinitely long process of mixing: can it at all be unmixed again even in principle?

This is the mathematical problem that Anaxagoras raises. He endows the Mind with the capacity to do the impossible. This is his most astounding scientific claim. (In modern mathematics one remains ambivalent on the issue of whether this claim can be made or not; see Sec. 1.5.) But before it was possible to address this central question, a simpler technical problem had to be solved first: How was the unmixing to be achieved?

At this point Anaxagoras introduces a technical term (his only one, mentioned 14 times in the different surviving fragments), "around-motion" (*peri-choresis*); cf. also Ref. 5a. He goes to great pains to make clear it is not a circular, a closed, a periodic motion which he has in mind. Birkhoff, who struggled later with the same problem, specially introduced his own technical term borrowed from the Latin: "re-currence" (around-motion).[6] Had he known of the older Greek term having precisely the same meaning, he probably would have adopted it.

From what has been said so far we see that Anaxagoras single-handedly created the qualitative mathematical notions used so successfully later by the Poincaré school: deterministic flow, surface-de-section (cross-section through a flow, i.e. recurrence) and — most important — the notions of mixing and unmixing.

1.4   A Machine to Illustrate the Idea

Here an important objection poses itself. Anaxagoras formulated his ideas in terms adopted from ordinary three-dimensional (or four-dimensional, with time included) intuition. Dynamical systems are defined in a space (the phase space) which is not intuitive and in principle can sport an unlimited number of dimensions (one for each variable). Can ordinary language and ordinary three-dimensional intuition at all capture the essence of the behavior of a nontrivial dynamical system in phase space?

The answer is yes. It only takes a "taffy-puller" and a rotating platform (a lazy Susan) in order to realize Anaxagoras' vision. The machine is illustrated in Fig. 1 (cf. Ref. 7). Taffy is that sticky and deformable sweet caramel mass so esteemed by the three nephews of Donald Duck. One finds the machine in the display windows of sweet shops in North America (like the Mall in Salt Lake City, where to the amazement of the surrounding children the first graphical notes were taken in 1978, and in Vancouver where Art Winfree stood transfixed two years later in front of a window when the same insight overcame him), but there may be many more places on the globe where the same machine can be found.

The taffy-puller is a mixing-generating machine — and at the same time the simplest example of chaos. One can picture it as being reduced to two dimensions. The curving arms [see Fig. 1(a)] make sure that there is an attracting symmetry plane present in the middle, so that the deformation of the taffy can be understood there as a two-dimensional process. The taffy-puller "is" therefore an area-preserving two-dimensional chaos-generating map (a Birkhoff diffeomorphism); see Fig. 1(c).

This machine can easily be placed off-center on a rotating platform in such a way that it stands transversal to the direction of rotation. The platform is driven by the same motor in synchrony with the arms. If now the lights are switched off, and a periodic (stroboscopic) flashing illumination of the same frequency is used, the taffy-puller can be seen at the same place again, only one rotation further, when the next flash comes. Alternatively a small fluorescent pearl can be put into the taffy. Then an ultraviolet light shows a complicated light path in the dark [Fig. 1(d)]. This experiment has yet to be carried out.

The thus-obtained nonrepetitive motion in three-dimensional space represents a nontrivial dynamical system. The three mathematical conditions which define a dynamical system are all fulfilled: trajectorial existence (through every point in the stroboscopically two-dimensional cross section there goes one light path), trajectorial uniqueness (through each such point), and continuous dependence on initial conditions of the trajectories (in the cross-section).

The system therefore is a "three-dimensional blender," as this class of dynamical systems has been called.[8]

Two additional remarks are in order. First, it is possible to take a video of the radiating pearl's trajectory in the dark and to then let the video run backward. The resulting motion is that of an "unmixing machine." Anaxagoras' dream can be implemented.

Second, it is possible to understand even better what goes on in three-space by building a simplified "analog model" under the assumption that the vertical height of the taffy approaches zero. To this end one needs only a sheet of

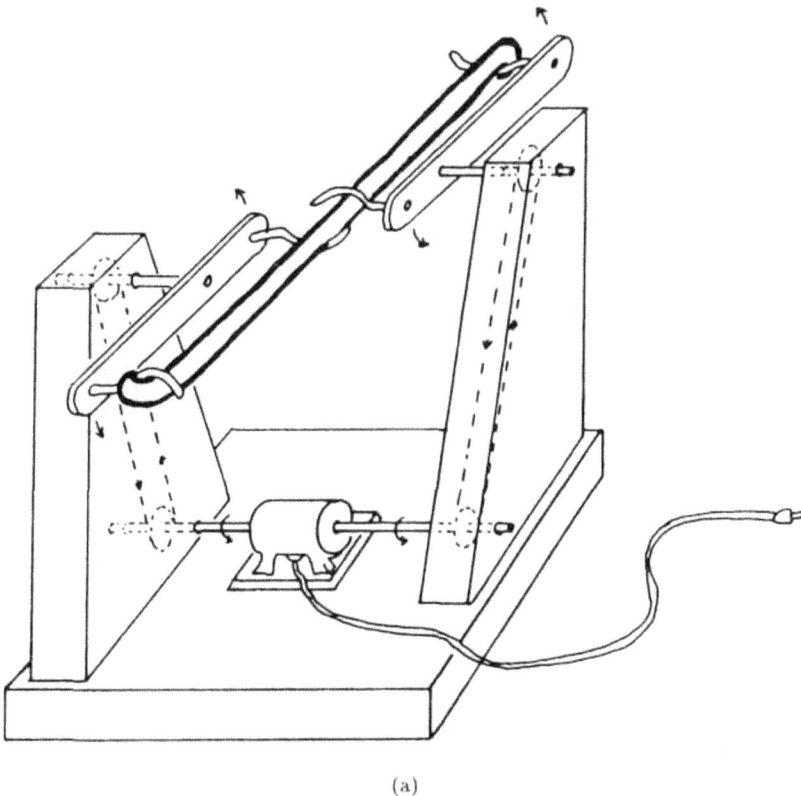

(a)

Fig. 1. A taffy-stretching machine ("taffy-puller"). (a) Schematic drawing. (b) Some successive positions of the arms, ranging from 0 to 360 degrees (shown in series). (c) An outline of the resulting two-dimensional map (a middle section through the taffy blown up vertically, before and after the first rotation). The dashing of the right hand portion of the taffy [in (b)] and the marker dots and gaps [in (c)] are meant to facilitate the identification of elements before and after the stretching. (d) The taffy-puller, placed on a corotating platform (including one glow point). (e) Simplified analog computer made of paper (cf. the text).

(b)

(c)

(d)

(e)

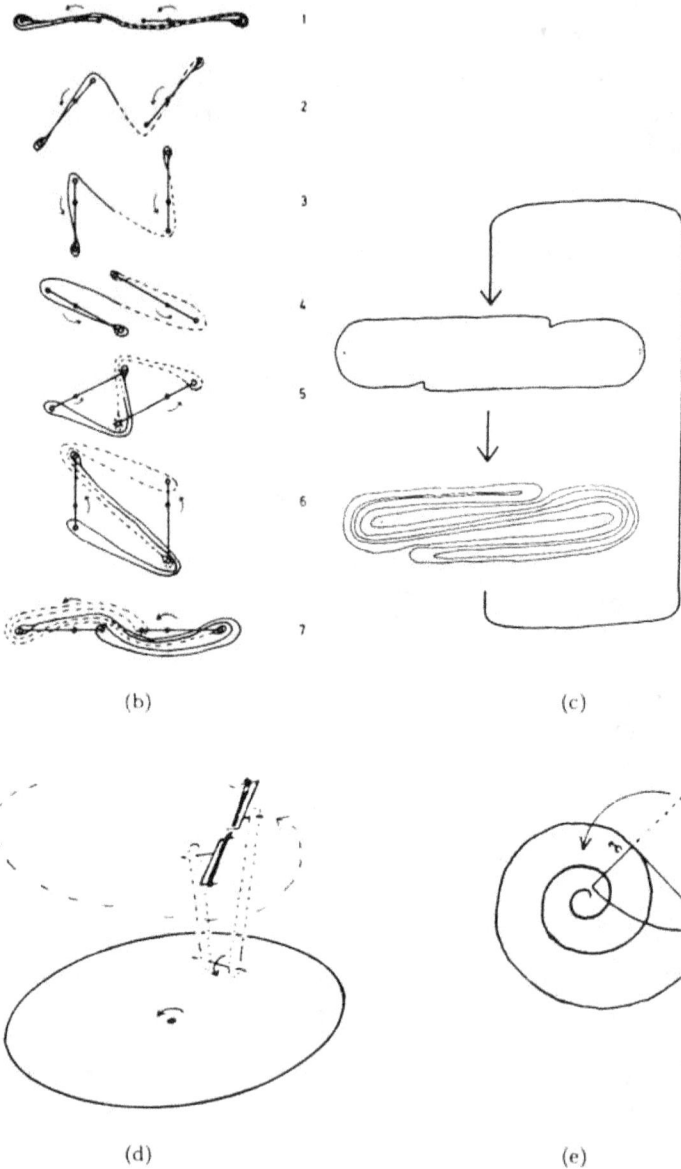

Fig. 1 (*continued*)

paper, a pencil, a pair of scissors and a glue-stick. First, an expanding spiral is drawn onto the sheet. Then, the spiral is cut out with the scissors in such a way that the inner turns remain intact while the outermost turn of the spiral is cut through in the direction of the midpoint until the last-but-one turn is reached. (The cut-out paper looks more or less like the letter G.) Finally, the outer flap is folded over toward the middle and glued on in this position by the glue-stick.

One can now use this little "machine" to trace out a path. The rotation is provided by the probing finger (or pencil, respectively). While one follows an inner curve on its outward path, an expansion takes place in the taffy. Then, when one comes into the outermost region, it looks as if one were moving into the range of an arm which first turns up and then moves down again toward the middle. Thereafter a new initial position in the interior part of the spiral is reached. That is, one has landed on a new spiral turn which one can again follow on its way outward, and so forth. Although this "taffy-puller" has only one arm and uses only one-dimensional (in a cross-section) taffy, the motion still exhibits all the chaotic properties of the original taffy-puller [cf. Fig. 1(e)]. One may after a while even hit on the old trick of using a stroboscope again. In this case, this means that one draws a crossing line somewhere and pays attention only to the subsequent crossing points generated by the moving pencil. One obtains a lawful dependence of the position of the respective next crossing point on that of its predecessor. The crossing line thereby becomes a "chaos-generating 1-D map," a one-dimensional recurrence.[8]

1.5   The Long Line

When looking at the sequentially stretched and folded-over dough (or taffy or 1-D taffy), or at the trajectories of an ODE doing the same thing,[7,8] one after a while begins to realize that something "impossible" is going on. It is the same feeling one may have had as a child when watching mother beat transparent egg white with a fork. I at least knew for sure (from my own unsuccessful attempts over a few rounds) that this procedure not only does not work but cannot possibly work. With every beat one just pulls the fluid a little bit around in a stretching motion. Certainly, no air can get trapped thereby beyond insignificant amounts — no matter how often one may repeat this purely symbolic motion. Nevertheless, lo and behold, as if there was a magic threshold this beautiful white foam arises.

The obvious reason for the perceived discrepancy consists in the fact that the mental model employed is additive — linear — while the process itself is multiplicative — nonlinear. Something exponential takes place here. More specifically, one could for example ask the question of what happens to the stretched material (the taffy or the fluid) in terms of its length. What happens to the length is, of course, nothing else than what happens to the width, only the other way around. Let us first look at the width. With each turn — each iteration — the width (vertical height) goes down by the same factor — of about 4 in the case of the taffy of Fig. 1(c). If a total rotation takes two seconds (as would be realistic), one has a factor of 2 per second in height change. Therefore, if one for example assumes that at the beginning the taffy on the left side consists of white molecules and the taffy on the right side of black (radioactive) molecules, then one needs exactly 24 seconds to achieve a reduction in height of both layers of 2^{24} or 10^8. Thus if the original height was 1 cm, one has after 24 seconds arrived at a height of 10^{-8} cm — the diameter of an atom. From then onward, maximal grayness has been obtained — a molecular sandwich of black and white molecules, as it were. Thereafter, one leaves all known physics behind with this exponential magnifying glass to enter the realm of pure mathematics. Here the process of probing the ever smaller never stops. Several remarks on the unlimited progression of smallness have survived in Anaxagoras' other fragments, where he claims that always a blowup to the old size is open to the Mind (cf. Fragments 1, 3 and 6 of Ref. 1, pp. 368, 370).

With the length it is the same thing the other way around. The original length — 1 m, say — is 2^n m after n seconds. This is huge after a short time already. What is of real interest from a mathematical point of view, however, is: What happens with the length when n is no longer finite? For only then can one speak of an invariant state, of a perfect mixture, of something that does not change any more over an infinitely long period of time. The small ("countable") infinity would need to be assumed for n (as Anaxagoras apparently already did). Then the obtained length is, of course, no longer finite either. But it also is not just infinitely many times larger than unity. Rather, it is 2 to the infinity, i.e. uncountably many times longer than unity (and infinity, for that matter). Thus it is a "long line" in the sense of mathematics.[9]

The long line of topology is a very unusual object. It is "nonseparable" (i.e. a countable infinity of points — like the periodic points — is no longer dense in it), and it is "noncompact."[10] Hence if one uses the natural topology of the line it is not an invariant set.[9] Anaxagoras' intuition — that such an object, the outcome of a perfect mixture, could still be "shrunk back uniquely" — is therefore vindicated (except for a set of measure zero[9]). On the other hand, if one uses the induced topology of the plane (as is standard usage in the early literature on chaotic attractors[11]), this result breaks down and there is no invertibility left after all finite transients have died out. The two mathematical alternatives are equally admissible. This result applies to attractors. A similar one holds true for area-preserving two-dimensional maps [like that of the taffy-puller, Fig. 1(c)]. This would mean that Anaxagoras indeed made a nontrivial claim.

1.6 Cloudlike Structures

A perfect mixture contains an unusual kind of curve — a long line — as we have seen. Are other, equally unusual kinds of object to be expected in the same context? Fractal curves, as defined by Mandelbrot,[12] are just as long as the long line, only that they are not smooth lines but rather contain uncountably many nondifferentiable points.[13] Intuitively speaking, one would therefore expect that such objects may likewise occur in the taffy. But what about the claim of unmixability in this case? We should not forget that we have only a locally parallel unmixing process available in accordance with the intuitive picture of a flow — a differentiable dynamical system as described above. Can an everywhere-smooth flow create a nowhere-differentiable fractal limiting object in a generic fashion?

If the answer were yes, the metaphor of the unmixable universe would even be capable of qualitatively explaining the overwhelming fractal aspect of nature as it is given to us empirically.[12] This fact could then be interpreted as a residue from earlier times — in accordance with Anaxagoras' claim (Fragment 12, Appendix) that the unmixing process has not been completely finished up to this day. This claim, by the way, inspired the young Kant to the modern theory of the formation of the solar system from a spherical cloud of gas. The latter condenses through "unmixing," first to form a spiral accretion disk and then to coalesce into the sun and the planets. Anaxagoras would also have liked the Hubble pictures of internally unmixing galaxies all over the place (including an occasional dark "supersun" and "superplanet").

An example which shows that indeed separating structures having the shape of a fractal nowhere-differentiable surface can generically arise in four-dimensional smooth mixing systems was recently given.[13] Unfortunately, the system in question is dissipative. Whether the same type of object can arise in volume-preserving flows of the type of Fig. 1, but of one more dimension, is currently open.

If so, this would mean that "cloudlike" objects are admitted into the perfect mixture and can be unraveled by a locally parallel unmixing process. Or, to put it in Anaxagoras' terms (Fragment No. 13,[1] p. 382), "The water got separated off from the clouds."

1.7   Conclusions

Smooth motions which are locally parallel everywhere cannot generate anything counterintuitive in two dimensions — as the study of well-combed two-dimensional hairdos revealed to Poincaré and Bendixson already. However, in three dimensions — and even in folded-up and glued-together two-dimensional flows (like that of the paper-and-pencil "analog computer" of Sec. 1.4) — such a "roundabout motion" suddenly acquires a multiplicative power. It generates a complexity which engulfs both the infinitely small and the infinitely large. The chaotic hierarchy (chaos in three dimensions, hyperchaos in four, etc.[7]) is an implication of this fact. In the fourth dimension, a qualitatively new phenomenon appears, giving rise to the "fractal hierarchy." It involves clouds in four and hyperclouds in five dimensions (cf. Ref. 14 for these notions). Could it be that the force of the "perichoretic principle" is so strong that with every new dimension (each additional variable), a qualitatively new type of complexity — with a new hierarchy — emerges?

This prediction at first sight appears easy to test. Like its predecessors, chaos and fractals, the new "X" phenomenon is probably thoroughly familiar to us from daily life. Nevertheless it may still take decades to be found if it indeed exists. The reason is that one just does not know what to look for this time.

Anaxagoras' idea of the mixing universe thus still proves almost as elusive to date as it did in his own time. Only the accent has shifted. While in former times the rewards of clear thinking were mostly philosophical, they are mostly applicational today.[15] Chaos and fractals are not only beautiful — the limits to their usefulness have yet to be defined. The world is full of systems with more than two variables and hence chaos, and full of systems with more than three variables and hence hyperchaos and fractals. Anaxagoras may have reckoned with the possibility of systems even with five or six — or almost infinitely many — variables. Is the Mind able to see even more in this machinery — does chaos give rise to an internal interface?

Acknowledgments

I thank George Marx, Esther Todt, Tamas Tel, Peter Erdi and Manfred Euler for stimulation. I also thank Art Winfree, Wolfgang Engelmann, Ed Lorenz, Stephen Smale, Ralph Abraham, Rob Shaw, Doyne Farmer, Norman Packard, Jim Crutchfield, Gottfried Mayer-Kress, Niels Birbaumer, Ilya Prigogine, Grégoire Nicolis, Masao Yamaguti, Kazuhisa Tomita, Yoshiki Kuramoto, Alexandre Ganoczy, Dick Noyes and Jack Hudson for discussions. György Targonski kindly drew my attention to the long line of topology.

References

1. G. S. Kirk and J. E. Raven, *The Presocratic Philosophers: A Critical History with a Selection of Texts* (Cambridge University Press, Cambridge, 1957), pp. 372–373.
2. Y. Kuramoto, personal communication, 1978.
3. A. Christie, *Chinese Mythology* (Paul Hamlyn, London, 1968), p. 44.
4. *Nova — Adventures in Science* (Addison-Wesley, New York, 1982), p. 27.
5. F. Lämmerli, *Vom Chaos zum Kosmos* ("From Chaos to Cosmos") (1962).
5a. A. Ganoczy, *Chaos, Zufall, Schöpfungsglaube — Die Chaostheorie als Herausforderung der Theologie* ("Chaos, Chance, Creation — Chaos Theory as a Challenge for Theology") (Matthias Grunewald Verlag, Mainz, 1995).
6. G. D. Birkhoff, "Surface transformations and their dynamical applications," *Acta Math.* **43**, 1–119 (1922).
7. O. E. Rossler, "The chaotic hierarchy," *Z. Naturforsch.* **38a**, 788–802 (1983). Reprinted in the festschrift *A Chaotic Hierarchy*, eds. G. Baier and M. Klein (World Scientific, Singapore, 1991), pp. 365–369.
8. O. E. Rossler, "Chaotic behavior in simple reaction systems," *Z. Naturforsch.* **31a**, 259–264 (1976).
9. O. E. Rossler, "Long line attractors," in *Iteration Theory and Its Functional Equations*, (eds. R. Liedl, L. Reich and G. Targonski), *Lect. Notes in Math.* **1163**, 149–161 (1985).
10. L. A. Steen and J. A. Seebach, *Counterexamples in Topology*, 2nd ed. (Springer-Verlag, New York, 1978), p. 71.
11. S. Smale, "Differentiable dynamical systems," *Bull. Am. Math. Soc.* **73**, 747–817 (1967).
12. B. Mandelbrot, *The Fractal Geometry of Nature* (Freeman, San Francisco, 1982).
13. O. E. Rossler, J. L. Hudson and M. Klein, "Chaotic forcing generates wrinkled boundary," *J. Phys. Chem.* **93**, 2858–2860 (1989).
14. B. Röhricht, W. Metzler, J. Parisi, J. Peinke, W. Beau and O. E. Rossler, "The classes of fractals," in *The Physics of Structure Formation: Theory and Simulation*, eds. W. Güttinger and G. Dangelmayer (Springer-Verlag, New York, 1987), pp. 275–281.
15. R. M. Goodwin, *Chaotic Economic Dynamics* (Clarendon, Oxford, 1990).

Appendix

Fragment No. 12 (On the Mind)

by Anaxagoras (*c.* 456 B.C.E.)

While all other things contain a portion of each thing, the Mind (NOYΣ) is without bounds and self-determined — mixing with nothing else but being alone by himself. For otherwise, were he not self-contained but mixed with anything else, he would be drawn into everything since in all things there is a portion of any other thing, as I mentioned. Whatever intermingled with him would prevent him from continuing to have power over everything as he has now, being alone by himself. Being the finest (lightest) and purest of all things, he has every knowledge about all things and possesses the greatest power. Also, all living things (things that have psyche), no matter how large or small, are controlled by the mind.

The mind assumed control over the whole roundabout motion (perichoresis, recurrence), setting it in motion in the first place. The roundabout motion first started out from the small, then more and more got drawn into it up till now, and even more will be involved in it eventually.

The mind knows exactly what is to be mixed together, what is to be kept separate, and what is to be divided off from each other. Whatever was meant to be — that which was but is not now, that which is now, and all that ever will be — got neatly laid out by the mind. This holds true also for the recurrence itself — as it extends presently to the stars and the sun and the moon as well as to the air and the aether, which are in the process of being separated from each other. It is the recurrence itself which is responsible for the separation process. From the rarefied, the dense gets separated off, from the cold the warm, from the dark the bright, and from the moist the dry.

Manifold moieties are there to all the many things. For all of them it holds true that none is completely separated, distinguished from the other — only the mind is. The mind is self-similar (totally similar) both in the large and in the small. Otherwise nothing is completely equal (similar) to any other thing. Always that which is most strongly represented in each thing determines most prominently its appearance.

2
How Chaotic Is the Universe?

Summary

The universe contains many chaotic subsystems, as well as many nonchaotic ones. However, there is a "maximum case" conceivable in which the whole universe becomes chaotic in the sense that every part is affected by a chaoslike perturbation. The simplest way in which this occurs is when the universe is "infected" by a chaos that is none of its own fault — because it is homemade, coming from the interior of the observer who watches the universe. This idea loses some of its outlandish character on closer inspection. Its roots go back to early findings made by Otto Sackur in 1911.

2.1 Introduction

The title of this chapter — courtesy of Arun Holden — promises more than anything that can possibly follow it. In the following, an answer will be presented which was not planned originally. At first, when nothing but the title existed, the plan was to start out with some remarks on Anaxagoras' invention of chaos as an explanation of the machinery of the universe — and on the transfinite subtlety of the only "immiscible" (nonchaotic?) substance, i.e. the mind. Then, the irregular "nailing" sound of a diesel engine, the lawful recurrence of Old Faithful the geyser, the behavior of X-ray bursters in the sky, and finally that of autonomous nerve equations, were to follow suit. An unpublished three-variable FitzHugh equation (whose chaotic analog computer solutions were shown to me by their late inventor in the Fall of 1976) was to round off the attempt to illustrate the ubiquity of "mixing processes" in simple nonlinear dynamical systems which populate the cosmos.

Yet, although these topics would perhaps not miss the challenge[1] (especially if not left without a remark on the problematical relation between continuous and discrete computational models[2]), something more down-to-earth will be attempted in the following: a "return to the mothers" of three-dimensional visualization.

2.2   Transfixed by Chaos

Let us look at a gas at equilibrium — a chaotic gas ("gas" is an artificial word meaning "chaos") of equal billiard balls. And feel the exhilaration of riding on such a ball like Baron Munchhausen (or inside it — equipped with transparent windows and a rubber coating, it makes a perfect bumping cart in three dimensions). Or just lean against one of the perfect walls of the container of the gas, in a safe little niche, and watch and listen.

It is like watching snowflakes fall. It takes a little while to get in tune and see the laws behind the whirling — like that I am moving up with the ground at constant speed (if I concentrate on the down-drifting balls). If the moving spheres are big enough and slow enough (and you are small enough in your niche to feel awed), you may suddenly "see" — at the risk of its turning out to be false on later scrutiny — that every ball owns a territory: one Nth of the volume of the whole container is assigned to it. And you "realize" that each ball is busy carrying out a duty — to contribute to the general pounding on the wall with the same vigor as all the others do on average. Even if its mass deviated widely from that of the other balls, the mean force exerted would still be the same. Yet if (as here) the masses of all particles are equal, the dutiful pounding job is even audible in principle (if not for the fact that, strictly speaking, no sound waves are permitted): the collisions with the wall occur at a fixed mean rate per unit area — it "snows" at a constant rate.

In other words, there is not only a unit volume [the mentioned Nth part of total volume (V/N)] present and hence a unit length (L), and a unit pressure (and hence unit energy E), but also a unit time (T) per unit length square. All of this has in principle been well known since the time of Waterston,[3] the first billiards aficionado in the history of physics.

Yet it is possible to step back even further, after this daydream, and re-member that what we were looking at — those huge floating spheres gliding by slowly and colliding silently and gracefully with each other while carefully observing a fixed mean density in space–time — was chaos. More precisely, it was hyperchaos (with the maximal number, $3N - 1$, of equal positive Lyapunov characteristic exponents, i.e. directions of repetitive stretching and folding-over in phase space[4]). This process is of an unfathomable complexity (see Ref. 5 for some of the mathematical details up to five balls) and produces a perfect mixture in the sense of Anaxagoras.[4]

Even if for the sake of simplicity a very special case were selected — a set of initial positions and velocities for which all the balls are forced to run on orthogonal tracks as if on a grid, interrupted only by their precisely coinciding collisions (so that the whole system becomes subtly converted into a discrete system, namely a three-dimensional cellular automaton of the reversible type in the sense of Fredkin[6]) — even then one could be sure already that the obtained motion possesses the maximum possible computational complexity. It would be beyond predictability for Gödelian reasons.[6] The above full chaos contains those special trajectories as infinitely rare unstable periodic solutions, embedded into the dominating "nonperiodic" ones (cf. Ref. 7).

2.3   Internal Chaos

Now we step back even further. Suppose you are in the possession of a classical Hamiltonian system of the above type, implemented in a desk-top computer of the firm "Laplace" which is so notoriously slow in delivering (its chips with infinitely many digits are still not available on the market). Or, even better, think of yourself as being implemented by a real-time machine of that kind. This is not totally inconceivable if one agrees to being gaseous — a "gaseous vertebrate" in the sense of Haeckel, maybe — since complicated dissipative structures including potential observers (excitable systems) are realizable on the basis of the Hamiltonian assumptions made.[8] And then ask yourself the same question again: How chaotic is the universe?

This time, it is not only a question of being (of how the universe or oneself is built) but also a question of appearing (of whether one's own identity can be recognized in principle). For example, if I were the only chaotic system in the universe, would not the whole universe appear chaotic to me? In other words, there is more than one sense in which the universe can be (or appear) chaotic. The less obvious second alternative is the one to be looked at in the following.

It might turn out, for example, that a universe which is chaotic ceases to be chaotic as soon as it is observed by an observer who is chaotic himself or herself. This possibility would lose some of its implausibility if one could show that observer-internal chaos is the source of quantum mechanics (which, as is well known, is chaos-free on the level of probability amplitudes).

2.4   An Action Implicit in Classical Chaos

In the following, a tiny step in the direction just outlined will be attempted. During the course of this attempt, only results already obtained in the above daydream will be used.

There is, as we saw in Sec. 2.2, both a characteristic "unit energy" (E) and a characteristic "unit time" (T) associated with every chaotic Hamiltonian system of the above type. In other words, since the product of a time and an energy is an action according to Leibniz, there exists a "unit action" (ET) for chaotic observers that belong to a certain classical type of universe.

This statement is, perhaps, not too surprising. The same result is already contained implicitly in the well-known Gibbs equilibrium entropy formula.[9] Gibbs found it in the wake of his celebrated "Gibbs paradox." The latter is worth a closer look.

2.5   The Gibbs Cell

Gibbs' recipe goes as follows. Take a certain volume of a gas (of the above type) and place it beside a second, equal volume of the same gas at the same temperature and pressure, but do not connect the two. Then you obtain a "state of mixing" of the combined system that does not change the mixing state of either of the two subvolumes as necessary. One expresses this by saying that the "entropy" (the mixing degree) of the combined volume is "twice" that of the original volume. It makes sense to define the mixing state in this fashion — so that the entropy "per volume" comes out a constant.

Now remove the barrier between the two volumes of the same gas. Each billiard ball can now roam about twice the former volume. The mixing degree therefore ought to have increased. But it has not: the total entropy (and entropy per volume) is *still* the same as before the barrier was removed. For if we close the barrier again, nothing has changed for any of the two gas volumes, as Gibbs correctly recognized as a young man of 36 in 1875. It is this fact which deserves the name "Gibbs paradox." For if entropy has anything to do with disorder and mixing, an increase ought to have occurred.

Usually, though, the Gibbs paradox is seen to lie in the following fact. Suppose the two gas volumes actually contain two slightly different kinds of billiard ball, but as much alike as you wish. In such a case, the new entropy per volume — after the mixing — becomes indeed much larger than before the removal of the barrier (namely, by a factor of $N \times \ln 2 \times$ Boltzmann's constant[10]). The phenomenon therefore has something to do with the distinguishability of particles.

In the present case, we stick to the above indistinguishable case where the entropy per volume does not change. Under this condition it is easy to continue. Suppose we have started out from just one of the two gas volumes, but now want to insert a barrier into its middle. What would be the entropy per volume in this case? Still the same. You guess the trend: no matter how busy we are, inserting (or removing) barriers between smaller and smaller subvolumes, the entropy per volume will always remain the same.

In this way, we even arrive at a numerical value for the whole entropy. The whole entropy is simply that of N volumes of the present gas at its present density — with just one particle left in each. We thus only need to know the entropy of a single particle, in one Nth of the original volume V — and we are finished.

Gibbs[9] conjectured that for this reason it suffices to insert, in the formula for entropy, just the phase space volume (the set of all attainable points in state space) of a single one-particle cell. (Since each cell has three dimensions, he added one more step for simplicity's sake, assuming that the final three-dimensional cell could be decomposed into three one-dimensional cells, each with its own one-dimensional particle contained.) And, lo and behold, everything came out fine. One so obtains the "equilibrium entropy" S:

$$S = k\,3N\ln P$$

In other words, the entropy S is proportional (with scaling constant k) to the logarithm (a nonlinear measure) of total phase space volume, P^{3N} (since all N cells, and even all $3N$ 1-D cells, are equal). The logarithm of the total phase space volume is then $3N \times \ln P$, the phase space volume of a single cell. The latter, finally, is

$$P = qp3.33\ldots ,$$

where q is the mentioned side length L of a cube of the unit volume V/N [namely, $q = (V/N)^{1/3}$] and p is the mean momentum of a particle per dimension [namely, $p = (mkt)^{1/2}$, where m is the particle's mass and t the temperature[11]]. The nonperiodic constant factor that starts with 3.33... is the hardest-to-explain part; it results from a lengthy calculation which involves the ratio of the surface to the volume of an N-dimensional sphere; a lucid description is to be found in Richard Becker's timeless book.[10]

The decisive point for us is that P, the unit phase space volume, is an action. This unit action is specific to the gas in question. It is, except for a factor close to unity, identical to the unit action (ET) which arose in the above case of the "floating spheres."

2.6 The Sackur Cell

It turns out that the Gibbs equilibrium entropy can be empirically measured. It then correctly describes the measured entropy of a gas at high temperatures.[12] The gas thereby behaves as if its molecules were billiard balls confined to phase space cells P that are never smaller than a certain empirically found unit action. The latter coincides (up to experimental accuracy) with Planck's constant.[12]

Alternatively put, division of P by an arbitrary fixed action (which makes S dimensionless, which is a desirable feature) gives a result which empirically yields a lower bound to naturally occurring values for P Sackur in this way empirically re-discovered Planck's constant in a gas, i.e. outside its original radiation context in a purely mechanical situation (cf. also Ref. 13).

Sackur saw that this mysterious constant in this way acquired a much better chance to be explicable from first principles.[14] He was aware that it appears that each gas possesses a characteristic unit action on classical grounds. While rediscovering the Gibbs entropy formula for a gas of indistinguishable particles, he also rediscovered the Gibbs unit cell (V/N). Thirdly, he noticed the existence of a unit time (the mean cell passage time, $2P/kt$) which goes with the unit cell.

After Sackur had died in a laboratory accident in 1914, his ideas were not forgotten. Ehrenfest and Trkal[15] were struck by his attempt to attribute reality to "cells" of size V/N. However, the Pauli cell, which was discovered only 12 years after Sackur's untimely death, has apparently never been linked to the Gibbs cell as rediscovered by Sackur. And the associated P, discovered and measured by Sackur, also apparently went unnoticed.

2.7   Conclusions

Macroscopic observers based on hyperchaos of the Sinai type are subject to the Gibbs–Sackur cellularity. What is the most important implication?

The most important implication can be seen only if one is gaseous. Only he who has chaos in himself can give birth to a rising star (said Nietzsche, who had read Anaxagoras). A chaos-generated cellularity may, if it is the finest systematic feature of a macroscopic system like you and me, leave some indelible mark on everything one touches or tries to touch.

The first who apparently saw this was John von Neumann, when he formulated: "The result of the measurement is indeterminate because the state of the observer before the measurement is not known exactly. It is conceivable that such a mechanism might function because the state of information of the observer regarding his own state could have absolute limitations, by the laws of nature."[16]

For homogeneous isothermal single-particle-type classical–mechanical observers, the Sackur cell P constitutes a limit to internal self-observation.[11] The argument used is similar to one introduced by Popper[17] to show that physical observers can never completely observe themselves. Popper derived from this the existence of an analogous limit to observer-external observation. If the Sackur cell is indeed the limit to self-observation, it may at the same time be a limit to external observation. That is, the internal unit action P of the observer may get imposed on the external world as an observational "uncertainty of action." The Sackur cell would then be the source of quantum mechanics.

The question of how an internally chaotic observer sees the universe has not yet been dealt with in an exhaustive fashion. Three general possibilities open themselves up. The first is that indeed P causes observations to be lawfully indeterminate, but that this "classical indeterminacy" pales before quantum mechanics — so as to be completely absorbed into it eventually. The second possibility is that everything falls apart, so to speak: there would be quantum mechanics and its nonlocality,[18] but there would also be other, equally nontrivial phenomena of classical origin — so that the interplay would become fairly complicated. The third possibility is monistic again — but with quantum mechanics no longer in the dominating position so that it becomes amenable to explanation.

Chaos itself is interesting enough as an object of study in its own right. Why should one, instead of studying chaos itself, turn the case around to make chaos an active participator in studying the properties of the world? Is there really a need for an analog of psychoanalysis in physics? If chaos were instrumental in bringing about a general shift of paradigm — away from the usual detached "exophysical" way of looking at the world toward an understanding, "endophysical" one[19] — this would only be another sign of the vigor of this surprising concept.

Acknowledgments

I thank Arun Holden, Michael Conrad, Hans Degn, Lars-Folke Olsen, Benno Hess, Joe Ford, Benoit Mandelbrot, Mitch Feigenbaum, Bob Rosen, Arnold Mandell, Evgeni Selkov, Anatol Zhabotinsky, Erik Mosekilde, Yoshisuke Ueda, Bruce Stewart, Bruce Clarke, Peter Kloeden, Roland Wais and Karl Haubold for discussions.

References

1. O. E. Rossler, "Different types of chaos in two simple differential equations," *Z. Naturforsch.* **31a**, 1664–1670 (1976).

2. M. Conrad and O. E. Rossler, "Example of a system which is computation universal but not effectively programmable," *Bull. Math. Biol.* **44**, 443–447 (1982).
3. J. J. Waterston, lecture read before the Royal Society (December 1845), reprinted in *The Collected Papers of John James Waterston*, ed. Lord Rayleigh (Edinburgh and London, 1928), p. 209.
4. O. E. Rossler, "The chaotic hierarchy," *Z. Naturforsch.* **38a**, 788–802 (1983); cf. also the previous chapter.
5. J. G. Sinai, appendix to English translation of: S. Krylov, *Works on the Foundations of Statistical Physics* (Princeton University Press, Princeton, 1980).
6. E. Fredkin, "Digital information mechanics," MIT preprint (March 1983); "Digital mechanics," *Physica* **D45**, 254–270 (1990).
7. T. Y. Li and J. A. Yorke, "Period three implies chaos," *Am. Math. Monthly* **82**, 985–992 (1975).
8. I. Prigogine, *Introduction to Thermodynamics of Irreversible Processes* (C. C. Thomas, Springfield, 1955).
9. J. W. Gibbs, *Elementary Principles in Statistical Mechanics* (Yale University Press, New Haven, 1902).
10. R. Becker, *Theory of Heat* (Springer-Verlag, Berlin, 1967).
11. O. E. Rossler, "Indistinguishability implies quantization," in: *Collected Papers Dedicated to Professor Kazuhisa Tomita: On the Occasion of His Retirement from Kyoto University*, eds. S. Takeno, T. Kawasaki and H. Tomita (Publication Office of *Progress of Theoretical Physics*, Kyoto, 1987), pp. 280–288.
12. O. Sackur, "*Die Anwendung der kinetischen Theorie der Gase auf chemische Probleme* (The application of the kinetic theory of gases to chemical problems)," *Ann. Phys.* **36**, 958–980 (1911).
13. H. Tetrode, "*Die chemische Konstante der Gase und das elementare Wirkungsquantum* (The chemical constant and the elementary quantum of action)," *Ann. Phys.* **39**, 434–442 (1912).
14. O. Sackur, "*Die universelle Bedeutung des sog. elementaren Wirkungsquantum* (The universal significance of the so-called elementary quantum of action)," *Ann. Phys.* **40**, 67–86 (1913).
15. P. Ehrenfest and V. Trkal, "*Ableitung des Dissoziationsgleichgewichtes aus der Quantumtheorie und darauf beruhende Berechnung der chemischen Konstanten* (Derivation of the dissociation equilibrium using quantum theory, and a thereby-arrived-at calculation of the chemical constant)," *Ann. Phys.* **65**, 609–628 (1921).
16. J. Von Neumann, *Mathematical Foundations of Quantum Mechanics* (Princeton University Press, Princeton, 1955), p. 439. First German edition, 1932: p. 233.
17. K. R. Popper, "Indeterminism in quantum physics and classical physics, Parts I and II," *Brit. J. Phil. Sci.* **1**, 117–133; 173–195 (1950). See p. 129.
18. J. S. Bell, "On the problem of hidden variables in quantum mechanics," *Rev. Mod. Phys.* **38**, 447–52 (1966).
19. D. Finkelstein (personal communication, 1983) suggested the use of these terms in place of "physics from without" and "physics from within," as had been suggested to him.

3
A Possible Explanation of Quantum Mechanics

Abstract

Any limit to observer self-knowledge implies an uncertainty of external observation. It is proposed that for "homogeneous" observers made up of only a few types of mathematically equal particles, the external world suffers a distortion. The underlying reason is that the observer is effectively replaced by an "observers' ensemble" since he does not know his own state exactly. Between the elements of the ensemble the observer jumps back and forth with a finite rate of "self-permutation." Since the elements of the ensemble are pairwise identical under time reversal, "micro-time-reversals" exist. The consequence is that all measuring chains become "effectively reversible" for the observer. Hence all external objects are subject to "pseudodiffusion." The latter resembles the diffusion postulated as an axiom in stochastic mechanics. This axiom (Nelson's axiom) implies a version of quantum mechanics, as is well known. In consequence a version of quantum mechanics can be derived from classical-deterministic first principles. The question arises of whether a "bootstrap principle" (fixed point theorem) is needed in addition, fixing the value of the diffusion coefficient and hence "Planck's constant" in a universal fashion. However, a fixed value of Planck's constant exists already — in the world of the observer. Other properties of quantum mechanics (like Wheeler's "delayed choice" and Bell's "nonlocality") appear to be implicit, too. Thus, a deterministic local explanation of quantum mechanics is accessible in principle under one condition: that a distinction between two levels of reality (exo and endo), one valid externally and the other only for the internal observer, is introduced. In this case an external reality can, paradoxically, be chosen again, which is not unfamiliar since J. Willard Gibbs made serious use of it almost a century ago. A new principle of parsimony of the world would thereby become visible: The "fancy" things would be true only on the endo level. Quantum mechanics would become an observer-specific endo reality.

Acknowledgments

I thank George Lasker, Jan Kryspin, Mike Mackey and Fritz Hund for discussions.

4
Endophysics — Physics from Within

Abstract

An objective physics must keep the observer at bay. The goal of keeping the observer outside can, paradoxically, be achieved only if the observer is explicitly included in a larger picture — which then is observer-independent again. When proceeding in this way, one realizes that the world is necessarily defined only on the interface between the observer and the rest of the universe. Since this interface is inaccessible as an object, there seems to be no solution left for internal observers like us. We cannot step out of our own world in order to adopt the role of a "superobserver." Hence we cannot understand the world.

Unexpectedly, there is a loophole: "model universes" can be set up which contain an explicit (microscopically specified) internal observer. As an example, it is possible to set up a classical molecular dynamics simulation of an excitable system which would act as a simplified "observer" in a larger artificial universe simulated in the same manner. One then, in addition to the observer, needs a "measuring apparatus" and an "object." The former could be a fluidic amplifier that acts like a power brake; the latter could be an ordinary particle in the universe in question that acts on the input piston of the pressure amplifier. (Three "reservoirs" would also be needed as an initial condition: "food" for the observer, a high pressure compartment to power the amplifier, and a set of "cool" particles allowing one to keep the first amplifier of the measuring chain sensitive for a while.) In such a scenario, the "interface" can be studied by us since we are on the outside. While it is still too early to do this explicitly in detail, it perhaps already makes sense to think about it and look at simplified cases.

We may for example assume that all micromotions in the observer fall into two equivalence classes — backward and forward motion — that are pairwise identical under time reversal. This would occur, for example, if the observer consisted of many equal pendulums of the same period. Such an idealization is admissible for starters. Then, while in reality it is the observer who alternates between black and white time slices — in the sense that every motion in the observer repeats itself in the opposite direction — to the observer it is the rest of the universe which alternates between being time-inverted or not. That is, relative to the observer, the external reality is subject to a succession of rapid causality reversals. This impression will be incorrigible.

The incorrigibility goes so far that even macroscopic records cannot correct for the "hallucinatory" impressions generated in the internal observer by the behavior of a micro-object. This is counterintuitive. At first sight there is no reason apparent why an "objectively presenting machinery" should not be able to overcome the limitation. However, a causality vacillation is "too deep" a property of the interface to be overridden by any auxiliary machinery in the universe in question. From this fact it follows that many objective features of the world (including macroscopic ones) which appear on the interface are only "observer-objective" but not really (exo) objective.

Collectively, these properties form a set undetectable to the observer since the fact that they exist only for him is nowhere represented. In the absence of a "tag" they represent what could be called the "Kafkaesque part of physics." A cure for this disability is impossible to obtain from the inside of the universe in question. Nevertheless the inhabitants can embark on an endophysical program of their own in principle — if we assume they are sophisticated enough on the macro level (consisting, for example, of many excitable systems as an advanced neuronal network). That is, they can build a still lower level universe themselves. Endophysics therefore becomes "level invariant" science.

In this way the inhabitants can begin to doubt the exo-objective nature of certain phenomena. In particular, "observer-centered" phenomena are likely to kindle their suspicion. Does it make sense in our own world to mark observer-centered phenomena with an "exophysical question mark" as well?

At least five types of phenomena in our own world are "observer-centered." They each arise from a privilege, viz.:

- Bohr's privilege
  ("Which reality am I to create by deciding to measure either the position or the momentum?")
- Heisenberg's privilege
  ("Where in the chain am I to place the cut?")
- Wheeler's privilege
  ("For how long am I to wait before choosing what to measure?")

- Bell's privilege
  ("Which nonlocal correlations at a distance am I to generate?")
- Everett's privilege
  ("What choice of world am I happy with?")

Analogs of all these suspicion-generating privileges exist in the model universe. The possibility of doing "meta-experiments" (as one would advise the inhabitants of the lower world to perform) therefore arises as an option for the inhabitants — as well as us. A glimpse of what lies behind the curtain is therefore accessible in principle, it appears.

Acknowledgments

I thank George Kampis and Peter Kugler for stimulation. For J. O. R.

5
Invention of the Name "Endophysics" — A Letter from David Finkelstein

Dear Otto:

Thanks for your kind words; and even more for feedback of any kind, which is rare.

I agree with your distinction between "inner" and "outer" physics, and am looking for good catchy names. "Exophysics" and "endophysics"?

But I am not really hopeful that the exophysics is classical, not quantum. Bell's inequality is the test.

Yours in the search,

David Finkelstein   June 23, 1983

6
Endophysics

Summary

A new science, endophysics, is introduced. Only if one is outside a nontrivial universe is a complete description of the latter possible — when you have it in your computer, for example. The laws that apply when you are an inside part are in general different: Endophysics is different from exophysics. Gödel's proof is the first example, in mathematics. In physics, it is desirable to have explicit observers included in the model world. Brain models are a case in point. Macroscopic brain models, however, are nonexplicit in general. Therefore an explicit microscopic universe is introduced. It is based on a one-dimensional Hamiltonian of the classical type in which "formal brains" can arise as explicit dissipative structures in the sense of Prigogine. The pertinent endophysics is still largely unknown. As a first step, the implications of the fact that the observer contains indistinguishable particles (Gibbs symmetry) are considered. Campbell's postulate — a fast back-and-forth vacillation of the time's axis on the micro level — is obtained as an implication. Nelson's postulate and hence the Schrödinger equation follow as corollaries. Thus, a "nonlocal" interface can be generated by a local theory. Microscopic observer properties "percolate up" to affect the macroscopic spatiotemporal appearance of the world. Physics becomes dependent on brain theory.

6.1 Introduction

Endophysics has so far been largely confined to science fiction. The best example is probably *Simulacron Three*, by Daniel F. Galouye,[1] which unfortunately did not find entry into Hofstadter and Dennett's excellent anthology on computer-cognition-relevant fiction.[2] Galouye lets a whole world arise as a computer simulation. The operator is able to look at this world through the eyes of the "ID units" — the poor inhabitants of that world. One inhabitant, code-numbered CNO (Zeno), unfortunately has to be programmed out because he gets suspicious and is about to infect the rest of the community. Only later does the evidence accrue to his creator, that he, too... but perhaps you wish

29

to read the novel yourself. (Eventually the two lovers, from different levels, live happily ever after since, after all, there is no basic difference between two subroutines which formally belong to two different levels of nesting.)

Shortly after Gödel[3] had presented his famous proof about the incompleteness (from the inside) of arithmetic, his close friend von Neumann[4] began to ponder the question of whether quantum mechanics might not represent an analogous limitation — within a physical rather than a mathematical context. Fortunately, von Neumann was able to prove that if quantum mechanics is accepted as the most basic physical theory which contains all possible others as special cases, there is no need to worry. For the structure of quantum mechanics happens to be such that "the state of information of the observer regarding his own state" cancels out from the formalism.[4] That this kind of "result" (Ref. 4, p. 439) is particularly prone to kindling suspicion in certain vulnerable individuals did not occur to von Neumann since he could not possibly have read Galouye.

About half a century before, a similar physical nightmare had already haunted Maxwell[5] (and apparently Loschmidt before him, according to Boltzmann[6]). Maxwell conjectured that there might in general exist two types of physical law. An example of the first kind would be Newton's law when applied to celestial bodies — it makes no difference whether or not you sit on one of the bodies. An example of the second kind would be Newton's law again, but applied to the many microscopic bodies whose mechanical interactions are thought to underlie thermodynamics. To be confined to that same world does make a significant difference. A being which belongs to that world cannot re-transform heat energy into mechanical energy without a temperature gradient. However, an external being (demon) can. Unexpectedly, this point of Maxwell's went unnoticed. The two famous proofs[7,8] that the demon cannot work (opening and shutting a little trapdoor of near-zero mass at the right moments in time) are valid only under the assumption that the demon is himself a partial system of the world in question. He then indeed cannot execute his task with a net gain. The fact that a much simpler mechanism than an intelligent being suffices in order to be just as efficient (or inefficient) was, by the way, overlooked at first. (An asymmetric trapdoor of near-zero mass needs only to be cooled from outside the world — with an infinitesimal amount of kinetic energy removed surreptitiously — in order to generate the desired effect automatically.[9]) The conclusion that the same effect cannot be obtained from the inside of the world in question nonetheless remains valid. (Operating a near-perfect cooling machine for a single particle will require the same investment of free energy from a demonic subsystem again as it can optimally gain from the cooled particle's "rectifying" operation.)

This limitation, however, applies to a being who is present *within* the system and hence is not qualified to be called a "demon." To a being who is not confined to the world in question, however, all of the above proofs break down. Indeed, when you sit at the keyboard of a computer in which a Hamiltonian universe is simulated, it is very easy to perform either of the above-mentioned magic tricks: Either you can adjust the tenth digit of a particular particle's position at a strategic point in time (Maxwell's magic trick, generalized), or you can make sure that a particular particle (the trapdoor) is very mildly and, from the interior of the world in question, unnoticeably cooled.

Hence the second law of thermodynamics is of an endophysical origin. Maxwell was right with his suspicion. The same conclusion holds good for Smoluchowski's[10] later proposal of an improved demon. He suggested a method for becoming a demon oneself: you only need — in modern terms — one of those highly sensitive infrared night glasses. The other requirements are trivial: a bowl of water, a plastic spoon, a dark room and two thermos bottles, one red and one blue. Then just wait and sample, with the spoon. That is, bright spots from the surface of the bowl go into the red bottle and dark spots into the blue bottle. Since any success would be sensational, there is no need for you to worry about the size of the effect. A thousandth of a centigrade will be fine; a fancy lab could be used later. Smoluchowski realized that if one is sure that even this tamed (macroscopic) version of the demon does not work, this conviction implies that one believes in the existence of counterintuitive nonlocal macroscopic correlations. The latter conspire to exactly "undo" the (from a macroscopic — for example, stochastic — point of view, unavoidable) success. Once more, one must have read Galouye to appreciate the depth of the fascination felt by Smoluchowski (before his untimely death from dysentery in 1917).

Next in line is Ehrenfest's demon — Einstein. In a letter written in 1927, Ehrenfest compared Einstein, more specifically, Einstein's untiring attempts to uncover a loophole in the consistency of quantum mechanics, with a "little devil-in-the-box" trying to play "perpetual-motion machine of the second kind in order to bust the 'inaccuracy relation.'"[11] Today, the thereby-initiated "quantum nonlocality"[12] has indeed acquired an analogous role as Smoluchowski's nonlocal correlations of thermodynamic origin possessed in the past.

Two further important names in the history of endophysics are Popper[13] and Finkelstein.[14] Popper talked (as he says in his autobiography[13]) Einstein into accepting his proposal[13] that complete self-observation is impossible in a deterministic continuous physics, and that one should try to find a Gödel type formulation of quantum mechanics. Finkelstein[14] designed a program for a "holistic" physics in the spirit of Bohr, but discrete. He hypothetically attributed both the quantum limit and the relativistic limit to the fact that the whole is not accessible to us. Later, he indicated an explicit example of a dissipative finite automaton (computer) whose internally evaluated state is different from the objectively existing one.[15] Shortly thereafter, he endorsed the two notions "physics from without" and "physics from within,"[16] by proposing the antonyms "exophysics" and "endophysics" as more attractive terms.[17] The name "endophysics" is his creation.

In the same year, Fredkin[18] described the first explicit, computer-simu-latable model universe — a cellular automaton of the reversible type. (Earlier "worlds" based on cellular automata — like Conway's game "life"[19] — had been irreversible.) This universe consists solely of information. As soon as you have realized it in any concrete form (whereby the most different kinds of hardware are conceivable), its properties are perfectly fixed: It starts to produce "material" properties of its own inside — like assemblies of hundreds of black pixels which stabilize at a certain size and then attract each other with a definite force law, such as Coulomb's.[18] Fredkin's hope is that some day, all laws of nature as we know them will follow as implications from such a reversible cellular automaton law. The only thing that is still wanting is luck — that the right reversible local rule will be hit upon. The number of rules to be checked empirically is unknown. Possible objections regarding the existence of nonlocal phenomena in quantum mechanics are countered with the argument that nonlocal correlations over large distances have been observed abundantly in computer runs — when such rules were calculated in real time on a fast parallel cellular automaton machine.[18]

Hereby the distinction between exophysics and endophysics is implicitly invoked. The same holds good for a seminal paper by Toffoli[18a] (Karl Svozil, personal communication, 1994). A problem with these beautiful explicit reversible model universes is that dissipative macroscopic processes cannot yet be obtained on their basis since even a single "elementary particle" consists of hundreds of cells already. Hence macroscopic-irreversible "observers" cannot yet be included.

These computer universes therefore belong in the first or "general" phase of endophysics. In this subdiscipline, one tries to find general limitations which invariably hold good on the inside. Gödel provided the paradigm and Maxwell the first potential example.

The second or "special" phase of endophysics, in contrast, cannot do without brain theory. Here, assumptions which are not of a completely general nature but are relevant only if an explicit observer ("brain") is a part of the model universe become essential. This makes the connection with Galouye's (and Stanislav Lem's[2]) science fiction even closer. The only difference would be that the present model universes are supposed to be microscopically rather than macroscopically simulated. In this context it is perhaps worth noting that the first potentially conscious computer program was developed in 1977 by Kosslyn and Shwartz[20] (see Ref. 21 for a more complete blueprint). This program is — like its predecessors of science fiction status — nonreversible. All such macroscopic models therefore wait to be imbedded into a more fine-grained (reversible) universe.

So far, a concrete example of a microscopically specified artificial universe which "goes all the way up" to include macroscopic subsystems as observers has been lacking. A universe of this type is introduced in the following.

6.2   A One-Dimensional Hamiltonian Model Universe

Endophysics stands or falls by a good strategy. If it is correct that the goal of physics is incomplete as long as the endophysical part is not included — since some empirical laws demonstrably lack an exophysical origin[5] — the task of physics is enriched by an added iteration loop.

In principle we cannot even begin, as long as we do not know our own exophysics completely, to develop the corresponding endophysics. Only Lady Luck (a correctly guessed rule in the sense of Fredkin) or some bootstrap principle[22] in the spirit of Baron Munchhausen could provide a shortcut. Endophysics therefore has to be worked out in a step-by-step manner — starting with very simple model universes. Once a model universe has been finished, the next — involving one or two more forces — has to be tackled, and so forth. Arithmetic is simple but some chains linking individually simple results are not.[3] This means that there is a danger of getting lost even during the first (zeroth) round of iteration.

An exophysics could — in accord with Mie,[23] who followed Helmholtz, J. J. Thomson and Bateman — be based on a field, i.e. a partial differential equation (PDE) of the reversible type. "Particles" (vortex rings or, in more modern parlance, solitons) would arise spontaneously from the field. This view is elegant and indeed presages the "local rule" in a reversible cellular automaton.[18] However, even the simplest classical PDEs are by definition "nonexplicit" in the sense that a good deal of their properties are still unknown. (As a witness, a classical relativistic electron was recently discovered.[24]) A PDE is like an animal species — you never finish learning something new about it (Ulrich Wais, personal communication, 1977). Therefore, it appears that one has to settle for a simpler universe first — an ODE (a system of ordinary differential equations) rather than a PDE. After this radical simplification, one might as well abandon relativity (an implication of Maxwell's PDE) with the same stroke. What remains as a candidate for a first continuous endophysics is the good old mechanistic billiard-ball universe of the 19th century.

This universe has gained new mathematical notoriety largely through the work of Yaakov Sinai.[25] He showed — even before chaos theory was the new wave — that Maxwell's notion of "weak causality" (which means nothing but sensitive dependence on initial conditions and hence chaos[26]) is indeed applicable to gases. Sinai obtained this result with the aid of a number of admissible simplifications: omission of one dimension (which leaves two); cutting down on the number of balls, first to two and then to one (by nailing the other onto the table); and eliminating all walls of the table (through identifying opposing sides). He was then able to prove that this "gas" possesses for almost all initial conditions exponential divergence between neighboring trajectories. Hence a classical billiard-ball universe is chaotic. A generalization of Sinai's result to more than two billiard disks and/or more than two dimensions is, nevertheless, exceedingly difficult since already in the case of just two billiard disks an eight-dimensional phase space and a four-dimensional configuration space have to be dealt with. Sinai[25] nevertheless gave several further results for systems with up to five disks.

If one wants to build up a whole universe, on the other hand, it is necessary to reach a level of geometric understanding which allows one to extrapolate from small to arbitrarily high particle numbers without effort. That is, a "building-blocks principle" is needed. Such a principle indeed exists — in one-dimensional billiard systems.

Perfectly one-dimensional systems — in a thread-shaped tube — would, of course, be dull. Shaking a tube with pills will hardly surprise you. So tricks are called for which keep the Hamiltonian one-dimensional but nevertheless enable new effects. Figure 1 illustrates the idea.

The system is not really "one-dimensional-straight" but is only locally one-dimensional. Nevertheless the Hamiltonian (the mechanical energy function from which all dynamical equations can be derived) remains one-dimensional. The one particle can only move up and down; the other only to the left and right. Nonetheless the interactions are energy-preserving, everything is locally smooth, etc. (see Ref. 9 for an explicit differentiable Hamiltonian). Although other classical chaos-generating Hamiltonians with only two degrees

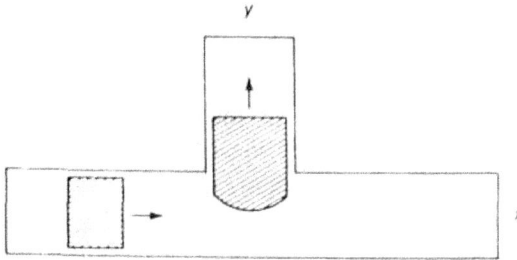

Fig. 1. A 1-D universe of two chaotic billiard balls.

of freedom are known,[27] the present system is specially easy to understand. In the limit in which the elastic walls of the frictionless particles and tubes become infinitely thin, the collision laws can be written down explicitly and one can calculate the trajectories in configuration space analytically (cf. Ref. 9), to obtain Fig. 2.

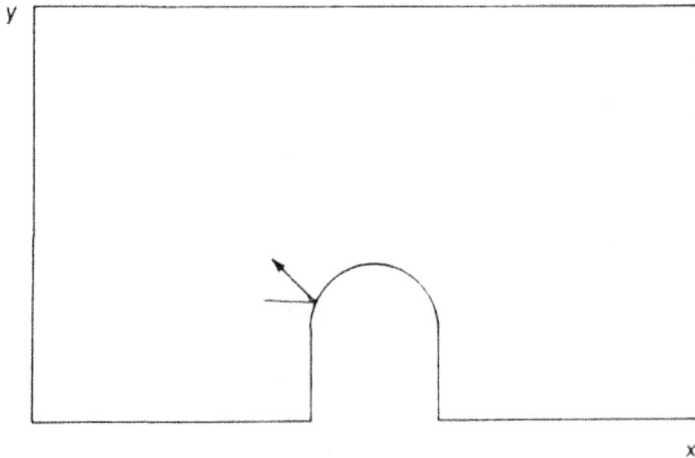

Fig. 2. Position space of the universe of Fig. 1.

The trajectory is that of a single, point-shaped billiard ball running on a two-dimensional billiard table. By Sinai's theorem,[25] there is sensitive dependence on the initial conditions (and hence chaos) whenever the trajectory is headed for the curved part of the protrusion. (No "canceling" effects are generated by the rest of the walls, so that almost all trajectories are affected.) Can this result be generalized to higher dimensions of configuration space? This is indeed the case. It suffices to insert a second (and third, etc.) horizontal particle so that they all interact only with the vertical particle. (This presupposes that they are transparent to each other, like solitons.) The Hamiltonian then remains simple. One obtains $n-1$ positive Lyapunov characteristic exponents (almost all equal).[28] Similarly, one is free to insert further vertical tubes containing a single particle each (or several that are mutually transparent). In all cases, one obtains a "maximally chaotic" ($n-1$ positive exponents) gas.[29]

At this point it becomes possible to redo statistical physics, so to speak, by checking numerous gas theorems obtained over more than a century against an explicit dynamical example (which makes a dream of Maxwell's[5] come true). But there is more to be gained: chemical reactions can be included. Elskens[30] recently looked at a (nonchaotic) one-dimensional gas containing several particle "colors" which upon meeting other colors in interactions were subject to lawful redefinition in accord with a quadratic mass-action kinetics. Only a "color look-up table" had to be added to the Hamiltonian. The same thing can be done here.

An especially simple explicit five-color mass-action system is available which generates a limit cycle oscillation as long as one particle type (color) is in large excess.[28] It can be imbedded in the above gas. The number of particles needed in either type of tube — vertical and horizontal — in order for the oscillation to become sufficiently "noise-free" can be determined numerically, for example by implementing this one-dimensional molecular dynamics simulation in a computer.[31]

Once a limit cycle (an irreversible macroscopic process of the nonlinear type which represents the simplest example of a recurrent dissipative structure in the sense of Prigogine[32]) has been successfully realized from first microscopic principles, so can other dissipative structures. In particular, an "excitable system" (a variant of a relaxation type limit cycle oscillator that produces only a "single-sweep oscillation" after a perturbation but becomes triggerable anew after a "refractory period") can be implemented. An excitable system, however, is a "formal neuron" (cf. Ref. 32). This means that a "formal observer" can be realized in the gas.

The explicit model observer can, of course, be made fancier. For example, a particle with a needle-like extension could be included, with the latter passing without friction through an outer "membrane" particle of the neuron so as to form a kind of piston. This does not appreciably increase the complexity of the Hamiltonian. As a second improvement, it will be desirable to have more than one formal neuron. In principle, were it not for present day computer limitations, one could opt for 10 billion neurons. In this way, virtually any formal brain can be realized in principle.

Apart from the observer, other complicated dissipative structures can be included in the simulation — like pressure amplifiers (of the power brake type), cooling devices and sensors involving wedge-shaped particles.[9] In this way, a whole reversible "complete artificial universe" can be generated involving objects, measuring chains and observers.

6.3 The Assumption of Indistinguishability

The artificial universe described above has yet to be implemented using a molecular dynamics simulation. Even the simplest case (a macroscopic limit cycle) is still extant.[31] It nevertheless appears safe to say that compared to the usual stochastic approach to dissipative structures, not many surprises will be in store. Macroscopic systems with familiar properties will be reobtained, and there will be fluctuations of a predictable size, depending on the number of particles used in the simulation. Thus, the conventional approach to artificial brains — using a macroscopic blueprint and a corresponding computer program — will be thoroughly confirmed and justified by the present explicit microscopic approach. It should be noted, however, that all these predictions are contingent on the exophysical perspective.

We now turn to the endophysics of the system. At first glance it appears impossible to make general predictions which will be valid across a large variety of formal brains. There is, however, one general property left which might harbor nontrivial implications. It is the assumption (made implicitly above) that particles of the same "color" always have the same mass and indeed do not differ from each other in any of their material properties — up to the point of being "unlabelable."

This is the assumption of "classical indistinguishability." Gibbs[33] already showed that this innocuous-appearing assumption drastically affects one particular macroscopic property of a classical gas: its phase space volume. The latter — and hence by implication the equilibrium entropy — gets dramatically reduced. His argument was short: the permutation invariance, induced by the indistinguishability (like in a game of cards involving N equal cards) renders $N!$ different states equivalent for any given state point in phase space. Therefore just divide phase space volume by $N!$. Later, Schrödinger[34] took up the challenge and declared that Gibbs had unwittingly made use of a quantum-mechanical result (namely, the invariance of the squared Schrödinger equation under particle permutation) without being entitled to do so.

This debate took a new turn more recently. Sudarshan and Mehra[35] opted for the position of Gibbs. The same holds good for Bach,[36,36a] who discovered that the quantum-statistical formalism itself — if done carefully — requires the introduction of the classical axiom of indistinguishability. In both classical and quantum statistics, the stochastic theory of "exchangeable random variables" (de Finetti's theorem) is the decisive tool. For example, this theorem taken alone apparently suffices for deriving the Bose–Einstein statistics.[36]

The possible existence of indistinguishable classical particles which are unlabelable indeed follows from soliton theory (cf. Ref. 37). What consequences does indistinguishability possess in a deterministic system?

6.4   Indistinguishability Has Endophysical Consequences

If we try to implement the above-described explicit model universe with the aid of a computer, the assumption of indistinguishability at first sight represents only an afterthought. This is because, in the computer, each variable has its well-defined name and its specific storage place. If particle indistinguishability possesses any nontrivial implications at all, they must be confined to the endo realm (or so it appears).

However, indistinguishability at first sight cannot exist endophysically either. If one was unsuccessful in attaching a label to a particular elephant which one wishes to monitor in the wild, one can still follow it with a helicopter and make an uninterrupted movie of its spatiotemporal trajectory within the herd. Such an argument was already given by Boltzmann — unlabelability can always be replaced by "spatiotemporal labeling."

Unexpectedly, this argument is not quite correct. If your camera team happens to have left unresolved a single situation in which a second elephant of the same size went into a crowded spot with the first and came out again, then for the rest of the expedition you will need two helicopters. The power of a name tag thus is unexpectedly strong. To make up for its absence-in-principle, only perfect observation is good enough (if at all). Perfect observation, however, is a problem endophysically. Popper[13] gave an early proof that complete observation is physically impossible. For example, ascertaining the microscopic trajectories of all equal-color particles contained in an observer is impossible on the basis of the macroscopic observational capabilities of the observer himself. A "second order observer" would be needed. Even the latter would be overtaxed, however, for chaos-related reasons. Only a "superobserver" would be up to the task. An "unresolvable indistinguishability" can therefore exist endophysically.

The limitation goes even further. To insist on a "complete" record (which goes back to the moment at which the initial conditions for the present trial run of the universe were set) is not only impractical from the inside, but it does not even make sense. For if the particles are perfectly equivalent as far as any of their individual properties are concerned, you need not know which is which — endophysically or exophysically. There is no situation envisionable in which exchanging one for its clone could affect the outcome of anything. This is a symmetry argument.

Perfect equality generates an invariance — invariance under a certain automorphism group of transformations.[38] In the present case this is the symmetric group.[38] Permuting the identities of the particles — that is, exchanging the label attached to the first pair of axes of state space with the label attached to the second pair, and so on — does not make for a "physically distinguishable situation."[38] This permutation invariance (and the whole present symmetry) was first seen by Gibbs[33] and used to obtain a quantitative result as mentioned.

6.5 Trajectorial Multiuniqueness

Gibbs assumed that to every state that the system may be in — a point in phase space — there exist $N!$ equivalent such points. This is correct. However, still more can be said. In general one is given only a single copy of the system. This system possesses a single trajectory of finite length (exophysically speaking). At every point in time, the system assumes a particular state point which lies on that single trajectory. Indistinguishability now means that one may take the two pertinent axes of each of the N particles and freely permute the labels over all those N pairs. (In a three-dimensional universe, one could interchange sixtuples of axes.) What one obtains in this way is many different versions of the given phase space that all look identical and contain the same unique trajectory inside: only the labels on the axes of the individual copies differ. A simpler way to look at this whole set of $N!$ equal specimens of phase space can be found, however — by "rearranging" the axes so that a single labeling scheme is valid again for the axes of all $N!$ copies. Whilst doing this, one must of course not change the values that are assumed by the given trajectory at any given moment along the axis in question. That is, while rearranging it is necessary to "pull along" the projections which the corresponding trajectory has on the axes. One in this way again obtains a single phase space

— in place of the $N!$ copies. It contains $N!$ mutually symmetric copies of the original (unique) trajectory. The so-obtained picture no longer changes no matter what system of labeling the axes is adopted.

Thus it is whole trajectories in phase space and not just points that are permutation-invariant ("multiunique"). More precisely, it is, at every single moment in time, $N!$ points which lie on so many trajectories. Phase space thereby becomes crowded. The unique trajectory which specified the system in the absence of permutation symmetry has been replaced by a multitude of $N!$ trajectories. The obtained tangle may be called a "multiunique trajectory." Particle indistinguishability thus implies trajectorial multiuniqueness.

Trajectorial multiuniqueness is a new mathematical notion, it appears. It can be introduced as a new axiom into the theory of ordinary differential equations, specifically those of Hamiltonian systems. What are its implications?

6.6   The Unit Cell in Phase Space

The new mathematical tool can be used to show that Sudarshan and Mehra[35] were right with their claim that particle indistinguishability implies the existence of an "irreducible representation of phase space." The "symmetrized Kronecker product" of the phase spaces of the individual particles whose existence they predicted exists. Moreover, it can be indicated quantitatively. Unexpectedly, the "unit cell" thereby obtained possesses well-defined "walls" and a unique internal trajectory.

In the present one-dimensional case, everything can be done by hand using simple pictures. Induction eventually leads to a general quantitative theorem applying to two- and three-dimensional Hamiltonian systems as well.[39] (Important steps in this direction have already been taken by Leinaas and Myrheim,[39a] in a paper quoted and reprinted by Witten.[39b])

We consider only the simplest possible case here, that of *two* equal particles in the horizontal tube of Fig. 1. (The vertical "third" particle may be either absent or retracted indefinitely into the upper tube.) Then the configuration space (the position of the first particle plotted versus the position of the second particle) has the properties seen in Fig. 3.

Both particles move back and forth between the one end of the tube and the other undisturbed so as if each were alone (which is possible since they were assumed to be able to pass right through each other). In this way, a quasiperiodic (Lissajoux type) trajectory is generated in configuration space. Only a very short segment — extending indefinitely into the future — has been depicted. On labeling the originally left hand particle

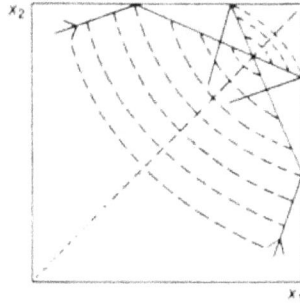

"particle 1," and the other "particle 2," one obtains the *upper* trajectory. One sees from the behavior of the upper trajectory that particle 1 moves through much of the tube while the second loiters near the right hand end, so to speak. However, if instead the originally right hand particle has been given the label "particle 1" and the other "particle 2," the very same behavior of the two particles leads to the *right hand* trajectory. The two paths are completely equivalent at every single moment in time. The momentary equivalence relations are indicated in the figure by dashed curved lines.

One also sees from the figure that the first bisector (the straight dashed identity line) plays a special role. The trajectorial segments that occur on either side of it are mirror images. One further sees that the identity line effectively acts as a "collision wall." Indeed, if a repulsive boundary had been entered in its place, an unchanged trajectorial picture would obtain on either side. Only a single continuous trajectory exists in each half-system — being reflected lawfully in an angle-preserving manner whenever it hits one of the three walls of its "cell."

This can be formalized. If we assume that the behavior of the two particles before the introduction of indistinguishability was described by a Hamiltonian H (see Ref. 28 for an explicit expression), it is now governed by the new Hamiltonian H':

$$H' = H + \frac{\varepsilon}{x_r - x_l}$$

That is, H' differs from the old Hamiltonian H by an added "collision term." The proportionality constant in front of it, ε, is assumed to approach zero. As a consequence, the particle on the left always remains particle 1 (or "l") and the particle on the right always particle 2 (or "r"). The two particles therefore behave qualitatively as if each had acquired a point-shaped "hard core" which interacts repulsively with that of the other. The new trajectories determined by H' differ from those determined by H only in added segments of infinitely short duration. The latter occur whenever a "switch" takes place from one of the two mirror-symmetric triangles toward the other along the straight dashed first bisector in Fig. 3 (at which the two particle positions coincide in real space). Then the two momentums of the two particles are instantaneously interchanged — whereby an infinitely fast "bridge segment" is generated in phase space. The latter, being visible in momentum space and whole phase space, cannot be seen in configuration space alone (since only the positions are plotted against each other in Fig. 3).

The question which poses itself reads: Is the "bridge segment" real? It would be real if it could be shown to be a mathematical implication of indistinguishability. To see whether or not this is the case, it is helpful to realize that the bridge segment coincides with one of the infinitely many curved identification lines whose existence followed from trajectorial permutation invariance. Such a "dashed connection" (Fig. 3) exists in the full phase space also at that moment in which the two particle positions coincide (so that the connecting line is no longer visible in configuration space but only in full phase space).

But why should this particular identification line be special? The answer is: It is not. It becomes especially visible only if one tries to decompose phase space into two identical portions. Then all identification lines disappear from the interior of the cells in question, because all interior points of a cell are identical to the same point in the next cell. Hence what remains to be understood is only the fate of those identification lines which do not cross a cell boundary. If such identification lines exist, a relation to the problem of the bridge segment is bound to make itself felt.

A more detailed look at the two triangles of Fig. 3 — and their extensions into full phase space — is thus required. All interior points match as we saw. But as we come closer and closer to the slanted cell boundary, the distance to the other fellow (in the other cell) eventually approaches zero in configuration space. One step before the boundary, the "match point" lies on the other side, one step beyond, on this side. What about the boundary itself? The match point then clearly also lies within the boundary.

If you picture the situation within the boundary in a blown up fashion, you realize that the boundary must be "double-layered." The one layer belongs to the one cell, the other to the other cell. Only under this condition can the two cells be identical in every respect. On the other hand, there must be no "gap" between the two layers since the recomposition of the two cells has to yield the full phase space again (no gap must remain). Each point on the skin of the one cell, and its match point on the skin of the other cell, therefore jointly represent a single point in the full phase space. Thus they represent what may be called "half-points." More precisely, they are "discretely adjacent points" in the sense of the topological theory of "local discreteness" of metric spaces in the sense of Bing's metrization theorem; cf. Ref. 40.

What are the properties of these half-points? First, we recall that each pair of full points (one within each cell) are, after the recomposition, again connected by an identification line of the dashed type. This line is thus a "two-way identification" (bijection). Second, in contrast to the two full points just considered, we now have four half-points to consider, two in the boundary of each cell in four-dimensional phase space. One pair is formed when the boundary is approached from the interior of the left hand cell of Fig. 3, the other when it is approached from the interior of the right hand cell. Note that in full phase space, the two identical cells which are present in a mirror-symmetric fashion in configuration space (Fig. 3), are no longer adjacent to each other in a mirror-symmetric fashion. A positive momentum corresponds to an equally large negative one, and vice versa. Hence the two approaching "light points" touch the boundary at four places as it were — two entry places and two exit places. (Think of a letter X made up of two brackets,)(, touching so as to form a letter I in the middle part, to be written with two hands simultaneously, the one hand starting on the upper right and the other on the lower left and each hand staying on its own side.) Each "pair" (the touching pair and the departing pair) has one representative in the skin of each cell. We therefore indeed have two half-points belonging to each layer. Since they are disjoint in general, they must be connected by a "one-way identification" within the skin of their own cell.

In this way, the full "two-way identification" is indeed reconstituted for all points in phase space once the two cells are recombined. For the two one-way connections now come to lie antiparallel to each other. This, however, is exactly the above-postulated situation of unidirectional intracellular identifications — the "bridge segments." Thus, what at first appeared to be a stumbling block has become a cornerstone. The irreducible unit cell indeed exists in a transfinitely exact fashion.

The two triangles (and their extensions into full phase space) therefore represent an irreducible representation of the biunique square of Fig. 3 (and its extension). At the same time, trajectorial uniqueness is restored within each cell. Since from time to time a one-way identification generates an intracellular "connecting jump" of zero duration, each cell is "multiple-connected" in effect.

The above Hamiltonian H' "simulates" this behavior with infinite — not transfinite — precision. There is an infinitesimal difference from the correct irreducible cell: the bridge segments in H' are full connections between full points rather than half-connections. In this way, two superfluous trajectorial (half) segments arise. Indeed, if two copies of H' with exchanged axis labels are combined, the full biunique phase space is reobtained — but this time with an infinitesimal gap in between the two cells. This infinitely small error is acceptable, however, if one is only interested in a simulation.

As an afterthought, it almost goes without saying that the above results are robust. With three (rather than two) equal horizontal particles given, the configuration space is a three-cube rather than the two-cube (square) of Fig. 3. The cube decomposes into six identical three-dimensional "triangles" (simplices) rather than two as before. Each of the six looks like a piece of black forest cherry cake (slightly out of shape but still bounded by planes).[29]

With N (rather than three) equal horizontal particles given, analogously configuration space becomes an N-cube. The latter decomposes into the "standard $N!$ triangulation of the N-cube" (as defined in Ref. 41). In this case, H' contains not one or two but $N - 1$ added collision terms.[28] Finally, also the admission of vertical particles (of the mutually transparent type, say) does not alter the partition of the full phase space in the above-described manner.

Even the main assumption of the above model — the one-dimensionality of the Hamiltonian — can be dropped and along with it the assumption of mutual transparency (noninteraction) of equal-tube particles. The "unit cell in phase space" can then still be indicated explicitly, it appears. To simulate it, however, H has to be combined directly with the pertinent rules for the relabeling of the axes, i.e. H' can no longer be written down in closed form. The necessary calculations may even turn out to be forbidding for large particle numbers due to NP-completeness, but this is a question for the future.

6.7   Main Implication

The main result is the existence of the pseudocollisions. The "permutation-induced boundaries" along which the multiuniquely populated phase space is "cut up" to yield an irreducible representation have the consequence that, compared to the prereduced description, an automatic "relabeling" of some pairs of axes takes place from time to time. This "self-permutation" is responsible for the fact that the variables valid inside the unique phase space cell are no longer identical with those pertinent in the symmetry-free description. The new unique trajectory — bridge segments included — refers to new variables. The old and the new variables, defined by H and H', respectively, are not identical even though they still represent mutual "analogs" in the sense of the theory of analogous systems.[42]

The spontaneous relabeling of the axes is not easily amenable to intuitive understanding. This is because it is the consequence of another counterintuitive mathematical property, trajectorial multiuniqueness. One has to pretend to believe in an automatic exchange of particle identities, at certain points in space, even though one knows that no actual physical permutation occurs. Of course, if such an exchange of identities occurred in reality, one would not be able to see it — so that it cannot be ruled out empirically. But even this excuse breaks down if more than one space dimension is considered. The real explanation, therefore, is that the physical assumption of perfectly equal particles is a "symmetry assumption." The counterintuitive implications (spontaneous permutation of identities at certain points in real space) are entailed by this symmetry. What are the consequences?

6.8 Quantitative Implications

First, Gibbs' "generic"[33] phase space volume is back. It differs from the nonreduced one by the factor $N!$ in the denominator. *Second*, "generic walls" apply which are generated by pseudocollisions in an explicit new "generic Hamiltonian" H'. It follows that it is possible to speak of a "generic cell." *Third*, a unique "generic trajectory" exists.

Fourth, a new point: there is a characteristic frequency, a "generic quasi-period" T. While the original system in the simplest case (only horizontal particles of the transparent type present) is quasiperiodic — with most particles reversing their direction of motion after a characteristic long mean period of time due to their having reached the opposite end of the tube — the generic system is quasiperiodic, too. However, the new mean quasiperiod T is distinguished from the old one by the factor N in the denominator. This is because the "compartment" to which each particle is now confined by virtue of the pseudocollisions is smaller by this factor than the original length of the tube, while the velocities are unchanged.

Fifth, an analogous characteristic period T is valid if the full original system (vertical particles no longer excluded) is considered. Although the original system is now no longer quasiperiodic but rather chaotic (exhibiting maximum hyperchaos[28]), the reduced system is quasiperiodic again. The reason for this is the compartmentalization. The pseudocollisions are far more frequent than genuine collisions (with a vertical particle).

In the simplest case of only one vertical particle present, the $n-1$ horizontal particles no longer acquire an equally large positive Lyapunov characteristic exponent each — even though in reality they all interact equally often with the vertical particle in a chaos-generating fashion, as we saw. The difference arises because only the innermost particles (being closest to the orifice of the vertical tube) have unimpeded access to the vertical particle. In those compartments, therefore, the frequency of chaos-generating interactions (and hence the magnitude of the positive Lyapunov characteristic exponent) is increased. Farther away, however, the values decline rapidly. So rapidly, in fact, that the fraction of "effectively nonzero" positive Lyapunov characteristic exponents soon approaches zero when the number of horizontal particles is increased (while the sum total of all positive exponents remains invariant).[29] Almost all new variables are therefore "effectively quasiperiodic."

The change in qualitative behavior comes about because the compartments to which the particles are now confined do — while "meandering" irregularly to the left and right — only rarely reach the middle of the horizontal tube where chaos is generated by collision with the vertical particle. This lucky event can occur only if all more right hand (say) particles have gathered beyond the middle. Hence the probability for the leftmost particle to interact with the vertical particle in the foreseeable future is virtually zero even with only 40 horizontal particles present (say). Thus one can understand that a systematic time-dependent "renaming" of the variables of a dynamical system suffices to change its qualitative behavior.

The same result — existence of a "generic quasiperiod T" — remains valid if the single vertical particle is replaced by many (mutually transparent, say) particles. Similarly if more than one vertical slot is allowed.

Sixth, in all cases the "quasiperiodized" motion reaches equilibrium much faster than this holds good for the underlying maximally hyperchaotic system.

Seventh, the introduction of color-changing reactions also does not interfere with the described quasiperiodization. The effect is again independent of whether the system is close to or far from equilibrium.

This is perhaps the most unexpected consequence of trajectorial multiuniqueness and hence of the Gibbs symmetry: that the irreducible representation of phase space and the related generic description are not tied to any equilibrium condition. Indistinguishability is independent of openness or closedness conditions. The same fact holds good for quasiperiodization.

6.9 Dissipative Structures

Eighth, dissipative structures that exist in the nonreduced Hamiltonian H remain in existence if the "generic" Hamiltonian H' is introduced instead. The latter seemingly generates the same pattern of criss-crossing trajectories in the x, t plane (all particle coordinates plotted versus time). However, there is a difference: equal-color lines now no longer genuinely cross each other. When one magnifies the seeming crossing points, they locally "avoid" the crossing. That is, instead of a "letter X," only the "pseudoletter)(" is present. Nevertheless, all concentrations of particles of a given color of course remain the same as before. Hence also a pre-existing limit cycle oscillation (for example) in the concentration variables is bound to be preserved.

Nevertheless, the compartmentalization (and the induced quasiperiodization) generates two further important effects: *Ninth*, large scale transport phenomena which existed in the nonreduced description of the dissipative structure in question cease to exist in their original form. For example, if "in reality" new green particles are continually generated on the left and, after having wandered to the right, disappear there in a lawful way (by acquiring a different color in a local collision, for example), the same situation looks different in the "reduced description": from time to time there arises a new green compartment on the left, and from time to time there disappears a green compartment on the right; that is all. All interjacent green compartments keep their identities while wandering very slowly to the right. In marked contrast to this slow wandering, however, a seeming "transport effect" sets in almost immediately (in accord with what occurs in the original description). This is possible because there is no identity relation any more between a green input and a green output particle. One could therefore speak of "transport without transport." In this way, at the same time a well-known quantum effect (transport of electrons through a metal wire) finds a classical analog and explanation.

Tenth, a further unexpected point: the whole dissipative structure "vacillates" rapidly in time with mean half-period T. In first approximation, it is almost as if the whole dissipative structure (which produces an autonomous temporal change of its own if it is a limit cycle oscillator) were subjected to a "shaking type" periodic forcing of very high frequency. On the time scale of this "fast hum," the macroscopic irreversibility completely disappears from view, leaving only the picture of a reversible process.

More specifically, what takes place here is only another consequence of the internal quasiperiodization. Every typical particle motion suffers a momentum reversal after a certain mean time interval T, as we saw. If we look at the x, t

plane again (because it is so intuitive), we can verify that most particles travel the other way within their own compartment, after a unit interval T has elapsed.

As a consequence, the intrinsic "microreversibility" of our far-from-equilibrium structure (cf. Ref. 32) is strongly enhanced. All the mediating and statistically averaging effects that the hyperchaotic mixing process exerts on the dissipative behavior on the nonreduced level, disappear, to leave us with a system that is both (pseudo)quasiperiodic and (pseudo)reversible and (pseudo)-equilibrated on a very short time scale.

The whole dissipative structure thus reverses its representative internal momentums (along with its mean momentum), after every unit time interval T. Equivalently, it is correct to say that the direction of time in which the dissipative structure is defined vacillates at a rapid rate (T). Hence a "microvacillation of time's arrow" is in charge for the dissipative structure in question.

6.10   Observers

At last, we reach our main object of study, the internal observer. The present approach was motivated by the aim to arrive at a special-endophysical result that would be valid for a "large" class of observers. Observers that (a) are dissipative structures and (b) contain indistinguishable particles do form a "large" subclass of possible internal observers in artificial Hamiltonian universes.

To what extent are such (in their own eyes macroscopic) systems affected by the details of their microscopic Hamiltonian? From experience with computers — which also constitute dissipative systems in almost all cases — one would expect some kind of "sealing-off principle" (the dynamics of the higher, more macroscopic level being sealed off from that of the lower, more microscopic one[43]) to be in charge. On the other hand, it has to be admitted that computers too (like brains) have never been analyzed "explicitly" on the micro level as this becomes possible to do in the present case.

The main finding of the preceding section ("quasiperiodization") remains valid if the dissipative structure in question is an observer. More specifically, we have: *Eleventh*, the direction of "most" particle motions inside the observer is reversed, after the mean time interval T, relative to all observer-external processes which occur in the model universe. The same fact when expressed somewhat more dramatically reads: the direction of observer-internal causality vacillates in relation to the observer-external direction of causality.

A similar claim has been introduced into physics once before — as a postulate. Norman Campbell[44] proposed in 1921 that all the phenomena characteristic of the quantum domain might be explicable by a single monistic assumption: that time ceases to be well-defined in the microrealm. More specifically, he said: "[Time is] a statistical conception, significant only with respect to large aggregates of atoms [so] that it is as meaningless to speak of the time interval between atomic events as of the temperature of an isolated molecule." This idea of time being "like temperature" apparently went into oblivion; Jammer's excellent book[45] at least makes no mention of it (even though two books by Campbell on a different topic are quoted). In the present context of an artificial simplified explicit model of physics, an analog of Campbell's postulate unexpectedly arises as an implication — that time's arrow vacillates on a microscale in generic Hamiltonian universes if they are observed from within.

The present situation has the asset that it can be studied at leisure since the phenomenon reduces to a geometric problem in two dimensions (the x, t plot). From kinetic theory it is known that any external micro-object which is directly coupled to the observer is subject to the unit thermal noise energy of the observer, E. This fact follows from Archimedes' principle of center-of-mass conservation, as is well known. However, the time reversals now cause this perturbation to be modified: *Twelfth*, the observer's unit thermal noise energy E is reapplied to a directly coupled particle, after every unit time interval T.

The consequence is rather inconspicuous at first sight, since a similar result would apply (with a formally different T) in its absence. *Thirteenth*, the directly coupled external object is subject to a well-defined diffusion law. Specifically, if M is the object's mass, the "effective diffusion coefficient" D is

$$D = \frac{ET}{M}$$

A simple statistical gas theory[46] and a more sophisticated theory (behavior of a diffusely reflecting particle put into a dilute gas[47]) lead to the same formula, which goes back to Einstein (cf. Ref. 48).

However, the effects of the micro-time-reversals of mean period T are thereby not exhausted. The black (causal) and the white (anticausal) time slices of the observer not only exert their effect on directly coupled external objects but invade the whole environment in a complicated fashion. Of special interest is the case in which the observer makes use of a measuring apparatus of the amplifying type. The measuring chain can actually be made arbitrarily fancy — involving a high gain, cooled sensors, etc., as described in Sec. 6.2.

The observer is only coupled to the "pointer" of the measuring device. The latter might, for example, consist of the large-mass "output piston" of the final pressure amplifier of the measuring chain. What the pointer does is determined by many earlier causal steps in the measuring chain and ultimately by the behavior of the measured micro-object. All these causal effects are reflected in the momentary behavior of the pointer. The pointer, however, vacillates between black and white, as we saw — between being in the one state of motion relative to all motions inside the observer and being in the opposite relative state of motion. All determinants of the pointer's momentary direction of motion therefore also reverse their influence on the observer when the latter turns white. This implies that the whole measuring chain oscillates between being black and white in synchrony with the observer, for the observer.

Since the measuring chain may be arbitrarily long (not only in space but also in time because of intrinsic time lags), it follows that the black and white states of the measuring chain are not confined to the objective simultaneity, but rather "bend over" across objective simultaneity in order to reach more or less deep into the past. This is, perhaps, the main result: *Fourteenth*, the observer is nonlocally coupled in space–time to his environment. In other words, a "nontrivial interface" connects the observer with his or her world, endophysically speaking.

6.11   Properties of the Interface

As it happens, the quantitative properties of the interface are easier to describe than its qualitative ones. Specifically, we have: *Fifteenth*, any indirect coupling is (pseudo)direct. That is, the two types of causal coupling between observer and object — causal and anticausal — applied in succession cancel out as far as the energetic side of the coupling is concerned. Hence what remains is a thermal coupling, reapplied every unit time interval, T. This sounds familiar. Indeed, we have: *Sixteenth*, the above diffusion coefficient ET/M remains valid in the case of an "indirect coupling" of the object to the observer, even if the coupling involves a measuring chain.

This can be seen more specifically. The amplified "push," exerted by the object on the observer in the black time slice, is followed in the white time slice by an equally strong "pull." In the negative white time slice, the observer seemingly causes the object through his own counterpressure (which in negative time is "disamplified" all the way back along the measuring chain) to perform the opposite motion at the front end of the measuring chain. Of course, the observer does not know about the two mechanisms (whether a black or a white time slice is in charge at the moment). All he or she notices is that the pointer is either pulled or pushed by the object — as if the object caused the pointer to do what it does "in response" to its own fine motions at the tip of the measuring chain — forward in the one time slice and backward in the other.

The resulting overall effect is therefore that of a reversible coupling. Causally vacillating observation chains are "effectively reversible." In contrast to an ordinary chain of levers which also acts in a reversible, energy- and action-preserving fashion, the present reversible causal chain is not strictly simultaneous but rather acts in a "time-bridging" fashion (a fact which possibly has testable implications[18a,b]).

The main consequence of the energetic (pseudo)reversibility is, however, that any object is (pseudo)directly coupled to the observer, as mentioned. The three parameters E, T and M hence "conspire" so that any external object in the world of the observer is subject to the (pseudo)diffusion coefficient D.

As an afterthought stemming from the real world, it is worth mentioning at this point that the above diffusion law is actually well known in physics — though not as a result but as a postulate. Nelson[49] showed that if this postulate is introduced as a single added axiom into an otherwise classical universe, it suffices to transform the latter into the world of quantum mechanics. In particular, validity of the Schrödinger equation is implicit (a fact apparently already seen by Schrödinger himself). Thus, if there has not been a mistake somewhere, quantum mechanics is a formal implication of a classical one-dimensional Hamiltonian universe — valid on the interface between an internal observer and the rest.

6.12 Conclusions

Endophysics is still in its infancy. So far, only a single explicit model universe is available that would reach through all levels, from the microscopic to the macroscopic. A general-endophysical result is the second law of thermodynamics, which according to Maxwell[5] is an endo implication of a classical universe. In the realm of special endophysics (including observers), the results of greatest interest are those that hold good for the largest number of observers. A large subclass of all possible classical observers is characterized by the presence of mathematical equality between particles of the same type that make them up. This is the only class that has been given closer consideration above.

Although, in the long run, particle indistinguishability may turn out to be only one among many endophysically relevant physical properties, some results of a broader significance appear to be implied — like a "parallelism" to certain known features of the quantum realm. Even more important from the point of view of endophysics, however, appears to be the following observation which was made along the way: simply putting a reversible universe into a computer and running it exophysically does not suffice to uncover its endophysics. In addition, one needs "hints" enabling one to look for endophysical properties even where exophysical correlates are absent. The endophysical significance of the Gibbs symmetry[33] (classical indistinguishability) is exophysically scarcely recognizable. Indeed, the need to take it into account in a molecular dynamics simulation has apparently never been felt up till now. Moreover, the two major implications of the Gibbs symmetry, i.e. quasiperiodization and micro-time-reversal, could have been overlooked easily as well — were it not for certain counterintuitive theoretical proposals already in the literature.[42,44]

The indistinguishability of particles has the further asset that it is a "maximally simple" property. Symmetries and reduced representations are staples of any physical theory. The property implied in the present case — trajectorial multiuniqueness — is, nevertheless, fairly nontrivial conceptually. The problem of when to trust a symmetry argument is still unsolved (cf. the discussion of Spinoza's "principle of the identity of the indiscernible" by Weyl[38]).

A more general endophysical question is the consistency problem. Can any endophysics be consistent? To what extent is "internal consistency" assured for its inhabitants when exophysical consistency is granted? Can an "internal interface" be consistent at all? How far does its consistency go, maximally? That is, are only single measurements covered (direct consistency), or are general laws derived from many measurements also consistent with everything else on the interface (indirect consistency)? How about "metaconsistency"? A metaconsistent world would be one in which it is impossible even to formulate an endophysical program (see Ref. 50 for a fictional account).

All of these questions can be studied explicitly using the present model universe, where the x, t plane represents the main conceptual tool. In the same way it will be possible to treat the question of "consistent interaction" between several observers, where the simplest special case is that of a single observer relying on his own notes, written on the occasion of an observation made earlier. The nontrivial nature of the same problem in quantum physics has been stressed by Bell.[51]

The central endophysical idea of meta-unmaskability goes back to Descartes.[52] He introduced the fairness question (in French): Can a *mauvaise plaisanterie* (a sadistic joke) be ruled out from the interior of the world? Both Einstein and Bohr agreed with Descartes that a physics whose consistency was not great enough to permit at least a glimpse at the reasons for our own limitations would be a "bad dream" (cf. Ref. 48).

In the present context, Cartesian fairness acquires a new facet. If only finite precision is available for simulating a Hamiltonian world in a computer, this flaw in precision is likely to destroy many "subtle" conservation laws. Subtle conservation laws would be those which preserve the consistency of internal interfaces. The second law, for example, is subtle since it can be violated by "late digits" manipulated by the superobserver.[9] Still more subtle is a macroscopically consistent world that contains nonlocal effects. Several mutually incompatible macroscopic worlds then coexist as implications of the same microscopic universe (exophysics). Only if such a degree of accuracy is guaranteed can the inhabitants hope to successfully embark on an endophysical program.

Therefore, an exact-reversible integration routine will be essential in the long run. (Such a computational universe is now available.[53,58]) Its use will come close to introducing a discrete "even-lower-level universe" lying underneath the above-discussed continuous one. Like Fredkin's cellular automaton universe,[18] this universe would then presumably be imbeddable again in an even-lower-level continuous universe (and so forth).

To conclude, endophysics is the study of demons. Maxwell's demon, Smoluchowski's demon, Gödel's demon and Ehrenfest's demon all do not work. They are each blocked by a censor. Further demons and their corresponding censors deserve to be uncovered. For to recognize and understand limitations is even more important than to be completely free of them.

Acknowledgments

I thank John Casti, Andreas Karlqvist, Per Sallstrom, John Wheeler, Joe Ford, Ray Kapral, Martin Hoffmann, Werner Lauterborn, Alan Lichtenberg, Ian Percival, Doyne Farmer and Mario Feingold for discussions. I also thank Edward Fredkin, Hans Diebner, Walter Nadler, Werner Pabst and Kurt Bräuer for subsequent discussions.

References

1. D. F. Galouye, *Simulacron Three* (1964). (A copy is available in the Dallas Public Library.) German translation: *Welt am Draht* ("A Puppeteer's World") (Goldmann, Munich, 1965); republished under the title *Simulacron-3* by Heyne, Munich, 1983 (No. 06/16, ISBN 3-453-30964-9). Cf. also the 1973 movie *Welt am Draht*, directed by R. W. Fassbinder (featuring Klaus Lowitsch and Eddie Constantine).

2. D. R. Hofstadter and D. C. Dennett, *The Mind's Eye* (Basic Books, New York, 1981).

3.   K. Gödel, *On Formally Undecidable Theorems* (Basic Books, New York, 1962). German original: *Monatshefte fur Mathematik und Physik* **38**, 173–181 (1931).

4.   J. von Neumann, *The Mathematical Foundations of Quantum Mechanics* (Princeton University Press, Princeton, 1955), p. 438f. German original: *Mathematische Grundlagen der Quantenmechanik* (Springer, Berlin, 1932, 1981), p. 233.

5.   J. C. Maxwell, *Theory of Heat* (Appleton, New York, 1872). Reprinted by American Mathematical Society Press, New York, 1972, p. 308.

6.   L. Boltzmann, "In memoriam Josef Loschmidt," in *Populäre Schriften* (Johann Ambrosius Barth, Leipzig, 1905), pp. 150–159 (in German).

7.   L. Szilard, "On the decrease of entropy in a thermodynamic system subject to interference by intelligent systems," *Behavioral Sci.* 1964, pp. 301–310. German original: *Zeitschrift fur Physik* **53**, 840–856 (1929).

8.   L. Brioullin, *Science and Information Theory* (Academic, New York, 1956).

9.   O. E. Rossler, "Macroscopic behavior in a simple Hamiltonian system," in *Lecture Notes in Physics*, Vol. 179 (Springer-Verlag, Berlin, 1983), pp. 67–77.

10.   M. von Smoluchowski, *Physikalische Zeitschrift* **17**, 557–585 (1916). See also his "Experimentally verifiable molecular phenomena that contradict the usual thermodynamics" (in German), *Physikalische Zeitschrift* **13**, 1069–1080 (1912).

11.   P. Ehrenfest, "Letter to Samuel Goudsmit, George Uhlenbeck and Gerhard Dieke," Nov. 3, 1927. Reprinted in *Niels Bohr*, eds. K. von Meyenn, K. Stolzenberg and R. U. Sexl (Vieweg, Braunschweig, 1985), p. 152.

12.   J. S. Bell, "On the Einstein–Podolsky–Rosen paradox," *Physics* **1**, 195–200 (1964).

13.   K. R. Popper, "Indeterminism in classical physics and quantum physics I," *Brit. J. Philos. Sci.* **1**, 117–133 (1950/51), p. 129. Cf. also his "Intellectual autobiography," in *The Philosophy of Karl Popper*, ed. P. A. Schilpp (Open Court, LaSalle, 1974), Vol. 1, p. 102f.

14.   D. Finkelstein, "Holistic methods in quantum logic," in *Quantum Theory and the Structures of Time and Space*, Vol. 3, eds. L. Castell, M. Drieschner and C. F. von Weizsäcker (Carl Hanser, Munich, 1979), pp. 37–60.

15.   D. Finkelstein and S. R. Finkelstein, "Computer interactivity simulates quantum complementarity," *Int. J. Theor. Phys.* **22**, 753–779 (1983).

16.   O. E. Rossler, "Chaos and chemistry," in *Nonlinear Phenomena in Chemical Dynamics*, eds. C. Vidal and A. Pacault (Springer-Verlag, New York, 1981), pp. 79–87.

17.   D. Finkelstein, "Letter to the author," dated June 23, 1983; cf. Chap. 5 of the present book.

18.   E. Fredkin, "Digital information mechanics," MIT preprint (1983); "Digital mechanics," *Physica* **D45**, 254–270 (1990).

18a.   T. Toffoli, "The role of the observer in uniform systems," in *Applied General Systems Research*, ed. G. Klir (Plenum, New York, 1978).

19.   M. Gardner, *Wheels, Life and Other Mathematical Amusements* (Freeman, San Francisco, 1983).

20. S. M. Kosslyn and S. P. Shwartz, "A simulation of visual imagery," *Cognitive Sci.* **1**, 265–295 (1977).

21. O. E. Rossler, "An artificial cognitive map system," *BioSystems* **13**, 203–209 (1981).

22. G. F. Chew, *Phys. Rev. Lett.* **9**, 233 (1962).

23. G. Mie, *Textbook of Electricity and Magnetism* (in German, 2nd edn., Stuttgart, 1912).

24. A. O. Barut, *Phys. Rev. Lett.* **53**, 2009 (1984).

25. Ya. G. Sinai, *Sov. Math. Dokl.* **4**, 1818 (1963); Appendix to the English edition of: S. Krylov, *Works on the Foundations of Statistical Physics* (Princeton University Press, Princeton, 1979).

26. U. Deker and H. Thomas, "The dice-tossing game of nature" (in German), *Bild der Wissenschaft* (Jan. 1983), pp. 63–75.

27. M. Hénon and C. Heiles, *Astron. J.* **69**, 73–79 (1964).

28. O. E. Rossler, "A chaotic 1-D gas — some implications," in *Proc. First Int. Conf. on the Physics of Phase Space* (College Park, May 1986) (eds. Y. S. Kim and W. W. Zachary), Lecture Notes in Physics (Springer-Verlag, Berlin), Vol. 278 (1987), pp. 9–11.

29. O. E. Rossler and M. Hoffmann, "Quasiperiodization in classical hyperchaos," *J. Comp. Chem.* **8**, 510–515 (1987).

30. Y. Elskens, "Microscopic derivation of a Markovian master equation," *J. Stat. Phys.* **37**, 673–695 (1984).

31. J. Brickmann and O. E. Rossler, "A recurrent dissipative structure derived from microscopic first principles" (in preparation).

32. G. Nicolis and I. Prigogine, *Self-Organization in Nonequilibrium Systems* (Wiley, New York, 1977).

33. J. W. Gibbs, *Elementary Principles in Statistical Mechanics* (Yale University Press, New Haven, 1902), Chap. 15.

34. E. Schrödinger, *Statistical Thermodynamics* (Cambridge University Press, Cambridge, 1946), Chap. 8.3.

35. E. C. G. Sudarshan and J. Mehra, "Classical statistical mechanics of identical particles and quantum effects," *Int. J. Theor. Phys.* **3**, 245–251 (1970).

36. A. Bach, "On the quantum properties of indistinguishable classical particles," *Lett. Nuovo Cimento* **43**, 383–387 (1985).

36a. A. Bach, *Indistinguishable Classical Particles* (Springer-Verlag, Berlin, 1997).

37. S. N. M. Ruijsenaars and H. Schneider, "A new class of integrable systems and its relation to solitons," *Ann. Phys. (N.Y.)*, 1986.

38. H. Weyl, *Symmetry* (Princeton University Press, Princeton, 1952).

39. O. E. Rossler and M. Hoffmann (in preparation).

39a. M. Leinaas and J. Myrheim, "On the theory of identical particles," *Il Nuovo Cimento* **B37**, 1–23 (1977).

39b. E. Witten, *Anyons* (Princeton University Press, Princeton, 1990).

40. J. R. Munkres, *Topology — A First Course* (Prentice Hall, Englewood-Cliffs, 1975), p. 222.

41. P. S. Mara, "Triangulations for the cube," *J. Combin. Theory* **20A**, 170–177 (1976).

42. R. Rosen, "On analogous systems," *Bull. Math. Biophys.* **30**, 481–492 (1968).

43. D. R. Hofstadter, *"Gödel, Escher, Bach — An Eternal Golden Braid* (Basic Books, New York, 1979).

44. N. Campbell, "Atomic structure," *Nature (London)* **107**, 170 (1921); "Time and chance," *Phil. Mag.* **I**, 1106–1117 (1926); *Nature* **119**, 779 (1927).

45. M. Jammer, *The Philosophy of Quantum Mechanics* (Wiley, New York, 1974).

46. F. Reif, *Fundamentals of Statistical Mechanics and Thermal Physics* (McGraw-Hill, New York, 1965), p. 486.

47. W. G. N. Slinn, S. F. Shen and R. M. Mazo, "A kinetic theory of diffusely reflecting Brownian particles," *J. Stat. Phys.* **2**, 251–260 (1970), p. 258.

48. A. Pais, *Subtle Is the Lord — The Science and the Life of Albert Einstein* (Oxford University Press, New York, 1982).

48a. O. E. Rossler, "Measuring the future," in *Measurement and Self-Similarity —* proc. Feb. 17–24, 1990 Zeinisjoch Conference (ed. P. J. Plath), in press.

48b. O. E. Rossler, *Das Flammenschwert — oder Wie hermetisch ist die Schnittstelle des Mikrokonstruktivismus?* ("The Flaming Sword — or How Hermetic Is the Interface in Microconstructivism?") (Benteli-Verlag, Berne, 1996).

49. E. Nelson, "Derivation of the Schrödinger equation from Newtonian mechanics," *Phys. Rev.* **150**, 1079–1085 (1961).

50. A. Strugatski and B. Strugatski, *Billions of Years Before the End of the World (Sa Milliard Let Do Kontsa Sweta)*, Snaniye-Sila (Sep.–Dec. 1976, Jan. 1977). German translation: Heyne, Munich, 1981.

51. J. S. Bell, "Quantum mechanics for cosmologists," in *Quantum Gravity 2* (eds. C. J. Isham, R. Penrose and D. W. Sciama) (Oxford University Press, Oxford, 1981), pp. 611–635.

52. R. Descartes, *Meditations on the First Philosophy (Meditationes de Prima Philosophia)* (Soly, Paris, 1641).

53. Edward Fredkin answered this "existence question" in the affirmative in June 1992, by jotting down on a table napkin (in Linz) a rule of how to set up an exactly reversible integration algorithm. The rule was shortly thereafter successfully implemented by Hans Diebner.[55] A class of algorithms which include the celebrated Verlet algorithm[56] as a special case was found to be implicit in Fredkin's hint (see also Ref. 54). A mathematical criterion specifying the conditions under which the algorithm remains physical was subsequently identified and implemented numerically.[57,58]

54. A closely related result was simultaneously and independently obtained by Levesque and Verlet: D. Levesque and L. Verlet, "Molecular dynamics and time reversibility," *J. Stat. Phys.* **72**, 519 (1993).

55. H. H. Diebner, "Investigations of exactly reversible algorithms for dynamics simulations" (in German), Diploma Thesis in Physics, University of Tübingen, 1993.

56. D. Verlet, "Computer experiment's on classical fluids," *Phys. Rev.* **159**, 98 (1967).

57. W. Nadler, H. H. Diebner and O. E. Rössler, "Space-discretized Verlet algorithm from a variational principle," *Z. Naturforsch.* **52a**, 585–587 (1997).

58. W. Pabst, "The principle of least action in discrete space–time" (in German), Diploma Thesis in Physics, University of Tübingen, 1995.

7
The Two Levels of Reality — "Exo" and "Endo"

Coauthor: Peter Weibel

Abstract

In the electronic age, the "interface" between observer and object becomes amenable to artificial manipulation. Perspective is, as is well known, not completely objective but only "observer-objective." Distorting the world is unavoidable if one is an observer, it appears. Relativity is the second example, and "mirages" are the third.

To set up a virtual reality with mirage-like properties is a modern challenge — an "interactive *trompe-d'oeil*," if you like. In the second author's institute, a touch screen and a touch-sensitive floor panel have been installed, with the aim of making the intimate entanglement between observer and environment perceptible as a both frightening and liberating experience. What is liberated is the mind.

The mind can be "made similar" to anything, according to Anaxagoras. The "Kon-Ton" (Hun Tun = mixture) of Chuang Tzu, the "Whole" (One) of Anaximander and the "XAOΣ" (mixture) of Anaxagoras are all early names for the intangible. The chaos in each case is accompanied by a twin notion: "Shu-Hu" (lightning) with Chuang Tzu, "Apokrinein" (a separating internal secretion) with Anaximander, and "NOYΣ" (Mindblade) with Anaxagoras. The dual notion in each case refers to the cut, the interface, the place where the cosmic egg pops open, respectively. It is a daring business to spring the bark of an internally burning log. But only if we catch the lightning by looking right into it (another ancient Chinese myth) can we glimpse the predivided (exo) world. The latter is still called "Kon-Ton" (chaos) in Japanese.

The first western scientist to recognize the interface problem was 18th century mathematician-cum-theologian Roger Joseph Boscovich, from Dalmatia. He asserted that the world really is deformable (we would say: "like rubber") but that we are unable to notice this. Due to our being made of rubber too, we are subject to codeformation. Hence the interface falsely presents a solid world to us.

Both quantum mechanics and relativity reflect this "cut type" (or endo) position. A mere observer-objectivity of the world ceases to be hermetic once it has been recognized. For it then becomes possible to investigate its properties with the aim of finding a loophole. A prison whose existence has been recognized can be sprung. Science and wisdom are no longer at odds.

Acknowledgments

O. E. R. thanks Yoshiki Kuramoto and the late Kazuhisa Tomita for stimulation. For J. O. R.

8
Explicit Observers

Summary

A new approach to the brain and the world is attempted. A chaotic Hamiltonian universe is set up in one dimension so that an internal observer — an excitable system — becomes amenable to complete understanding. The Gibbs symmetry and the Wigner symmetry, when taken into consideration explicitly, imply that to this observer, the world appears different from what holds good objectively — such as if one is doing a molecular dynamics simulation of the same system, and looking at it from the outside. Specifically, both stochastic mechanics and the quantum nonlocality turn out to be formal implications of the present artificial universe — which thereby qualifies as a local-deterministic hidden-variables approach to quantum mechanics at the same time. Bell's well-known impossibility theorem is circumvented because all quantum effects arising are nonexistent objectively. They are valid only within the "interface" that develops internally between the observer and the rest of the universe. Hence for the first time the Kantian notion that the world is objectively different from the way we perceive it can be demonstrated — not for our own world but for a lower level universe as it presents itself to an artificial observer living inside.

8.1 Introduction

"Artificial life" is an approach to evolution and the brain.[24,26] As such it is explicit in the sense that the structures in question can be implemented in the laboratory and/or the computer. However, the pertinent equations are in general dissipative. This means that an even "more explicit" description is possible in principle, in which the irreversible equations are replaced by an underlying set of reversible ones.

An example is Fredkin and Toffoli's discovery that any irreversible cellular automaton (CA) can be embedded into a larger reversible one.[10] Only the latter would be "maximally explicit." Similarly, the ordinary or partial differential equations (ODEs, PDEs) describing an evolutionary soup[23,24] or a system of well-stirred[25] or non-well-stirred[35] neurons (cf. the simulation of an excitable PDE with chaotic behavior[35]) are necessarily dissipative. In these cases an underlying, maximally explicit description contains, not twice as many variables as in the CA case,[10] but 10^3–10^{23} as many. A fully explicit description would be a molecular dynamics simulation (MDS)[1] in which the dissipative "macroscopic" behavior is the outgrowth of of a reversible microdynamics of vastly more Hamiltonian variables.

The same macro/micro distinction applies also to virtually every computer built so far. The discrete finite state description is nothing but a shorthand that captures certain invariant features of an underlying continuous system of ODE's that is "decomposable" in the mathematical sense.[27] This latter system in turn belongs to the class of dissipative ODEs already discussed.[25] Thus every computer-implemented observer in our own world will once again require an underlying Hamiltonian description on the micro level.

Ultimately, of course, any such description would, in our own world, have to be replaced by a description in terms of the Schrödinger equation as a first step on the way to a final quantum-field-theoretic description hopefully available in the future.

The advantage of the artificial life approach, however, is precisely that nothing forces one to choose an explicit universe that is "completely realistic" at the same time. Thus, artificial observers that arise in a reversible cellular automaton are of the same standing, in principle, as are artificial observers arising in a molecular dynamics simulation, and as will be artificial observers generated in a "quantum molecular dynamics simulation" of the future.

In the following, the second category among these (the currently accessible MDS) will be considered. All results found will nevertheless apply equally to a lattice automaton analog (first category).

8.2 Explicit Dissipative Structures

The observers to be simulated belong to the class of "dissipative structures" (or, synonymously, far-from-equilibrium structures) which figure so prominently in the work of Prigogine and Nicolis.[20] Two decades ago the prevalent attitude among biochemists was that going down to this "next-lower" level of description would almost never be necessary when dealing with a concrete system like the oscillatory Zhabotinsky reaction. Certainly, you would have to make sure that your macroscopic "rate equations" are open in the sense of containing "pool substances" that are kept either constant (so that the system can develop a far-from-equilibrium attractor) or initially far enough away from their final equilibrium values that an attractor-like transient dynamics can develop. This is the case when the reagents of the Belousov Zhabotinsky reaction have been freshly poured into a beaker, for example (analogous to a portable radio that has been equipped with a fresh battery). Even the principle of detailed balance[41] which is ultimately of microscopic origin can be incorporated into the macroscopic rate equations, directly.

Of course, it would be "desirable" to include "fluctuation terms" in the deterministic rate equations under certain conditions — as in the presence of a bifurcation or phase transition — when capture of the details of the amplified fluctuations is at stake, for example. "In general," however, knowledge that the need for a second order approximation may arise would suffice.

Until recently there was perhaps indeed no need to bother. Only physico-chemical conscientiousness — and some technical applications — seemed to justify research programs that from the beginning included the next-lower phenomenological level.[20] From the point of view of scientific efficacy — when the finding of new phenomena like chaotic behavior on the macro level was at stake, for example[28] — it would even be a mistake to keep the next-lower level of description in mind continually. The hierarchy of sciences — quantum field theory at the bottom, the behavioral sciences at the top — fortunately works most of the time. It would be a major catastrophe if it turned out that to understand the brain, for example, you would have to use the quantum level of description.

The chaos paradigm changed this attitude. Suddenly, there was a dynamical principle which (a) was very powerful in terms of the predictions it made — including the presence of an unsurmountable horizon of unpredictability[15a,39]

and (b) was independent of the level on which it was applied. For example, populations of butterflies that each employ a chaos-generator in their individual flight maneuvers could easily behave chaotically, too. Of course, the mere presence of analogies between levels does not yet mean that the levels themselves have been punctured. Nevertheless chaos theory for the first time offered a unified approach to random-appearing events on all levels. Should, for example, the irreducibly stochastic behavior found on one of those levels — the quantum level — really prove resistant to the new paradigm?

Looked at in the light of chaos, the Boltzmann–Prigogine research program of going directly to the "correct" micro level of description when phenomena like entropy increase and evolution are at stake, suddenly made eminent sense. After all, no one had yet fully understood how reversible trajectories generate an irreversible macrodynamic. Chaos theory, applied to Hamiltonian systems with a few degrees of freedom, might be able to reveal the ultimate mechanism.[30] Indeed, the momentary phase space volume appears to represent an explanatory "deterministic entropy" (cf. Refs. 30, 5d).

While the reversibility paradox thus appeared in a new light, should one really believe that this paradox might have anything to do with quantum mechanics, or with the way an observer perceives his or her world? Moreover, would this latter possibility not invite back the catastrophe of level mixing?

The answer is yes. Nevertheless, chaos demands that one try. All that is required is having another look, after Boltzmann and Gibbs, at what "really" happens on the micro level. The difference would be that, while Boltzmann and Gibbs thought they were doing physics, this would be nothing but chaos theory today.

At this point, everything becomes very simple. There exists this molecular dynamics simulations paradigm[1,4] which, it turns out, is "pure chaos" from the beginning (as Maxwell had known). It is a universe to explore.

8.3   Molecular-Dynamics-Simulation-Borne Observers

Dissipative structures have been simulated before from first reversible principles in a molecular dynamics simulation; the example of bistability (more precisely, autocatalysis) was studied in this way over a short time using a partially stochastic algorithm.[13] However, no recurrent dissipative structure — an analog of the Belousov–Zhabotinsky reaction (with its autonomous oscillatory and/or chaotic[14] and excitable[35] behavior) — has apparently been simulated explicitly from deterministic reversible continuous first principles up till now.[5e]

The reason for this lack of success is simple. The number of particles needed to get a working simulation in which the autonomous macroscopic behavior (generated on the basis of differing rate constants) is not obliterated by noise, in two or three space dimensions, is prohibitively high even for present day computer power.

Therefore a return to the chaos-theoretic approach is called for. Indeed it turns out that no more than a single space dimension is needed. An example is sketched in Fig. 1.

Here we have two loaf-shaped particles living in a frictionless 1-D tube. An appropriate Hamiltonian (which can be transformed into a set of differential equations that can be put into a computer) is

$$H = \sum_{i=1}^{n} \frac{p_i^2}{2m} + \varepsilon \sum_{i=1}^{n} \frac{1}{x_i} + \varepsilon \overset{n,n-1}{\underset{i,j=1,2}{P}} \frac{1}{1 - f(x_i) - x_j}. \tag{1}$$

Here the symbol P refers to all possible permutations between particles (since all can interact). The masses are all assumed equal ($m = 1$). The particle number assumed is n (for example, $n = 2$). The small parameter ε determines the "hardness" of the particles' walls. The limit of rigid elastic reflection is reached when ε approaches zero (more precisely, when the ratio of ε and H, the constant total energy allotted to the system, for example, $H = 1$, approaches zero). The function f, finally, may for example have the form of a "blunted tent in the prairie viewed against a bright evening horizon":

$$f = \sqrt{(x - 0.05)^2 + 5 \times 10^{-6}} + \sqrt{(x - 0.15)^2 + 5 \times 10^{-6}}$$
$$- 2 \cdot \sqrt{(x - 0.1)^2 + 10^{-4}}. \tag{1a}$$

(Cf. Refs. 30, 31, 33 and 36 for some alternative functions for similar or related geometries.)

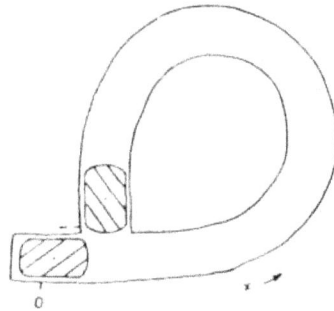

Fig. 1 A chaotic 1-D Hamiltonian system.

The system of Fig. 1 with two particles [described by Eq. (1) with $n = 2$] is chaotic. This becomes obvious from Fig. 2, which shows the position space (also called configuration space) for this system as ε approaches zero. All possible particle positions of the first particle, x_1, are plotted against all possible particle positions of the second particle, x_2, whereby two trajectories are symbolically indicated.

One sees that the system is equivalent to a 2-D chaotic Sinai gas.[40] Specifically, neighboring trajectories — impinging in parallel on one of the two curved protrusions in Fig. 2 — will diverge and will keep doing so exponentially.[40] Thus, there is a positive Lyapunov characteristic exponent — implying chaos — present in this 1-D Hamiltonian system.

A point-shaped billiard ball running on a frictionless table is indeed equivalent to a two-particle one-dimensional billiard game.[4a] (Joseph Ford, personal communication, 1986). If the system is not "hard" but "soft" (i.e. ε is greater than zero), there arise, in addition to the Sinai chaos, "nonchaotic islands" in phase space whose relative volume can be made arbitrarily small by decreasing ε.

A second result is that the present system can be generalized to arbitrarily high particle numbers (and hence dimension numbers of configuration space), without losing its mathematically transparent character. With three particles, for example, we obtain Fig. 3.

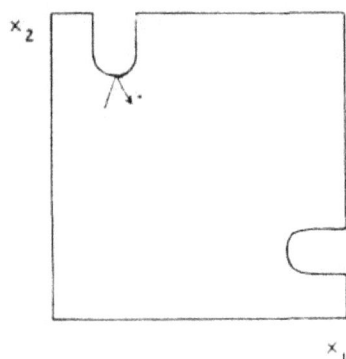

Fig. 2. Configuration space for the system of Fig. 1 with two particles (schematic drawing). A computer simulation of Eq. (1) with $n = 2$ and $\cdot = 0.0001$ and $H = 1$ yields a similar picture.

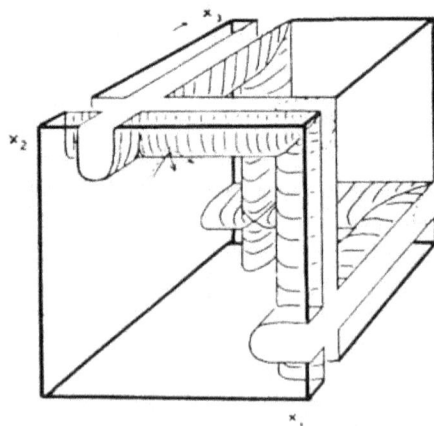

Fig. 3. Configuration space for the system of Fig. 1 with three particles, corresponding to Eq. (1) with $n = 3$. Two neighboring trajectories are indicated again (cf. Fig. 2).

There are now two different projections (or representative cuts, respectively) present which each look like Fig. 2. The corresponding Hamiltonian is again Eq. (1), but with $n = 3$ rather than $n = 2$. This "building-blocks principle" continues. For n particles, we have $n - 1$ equal projections of phase space each looking like Fig. 2. It follows that the system generally possesses $n - 1$ positive Lyapunov characteristic exponents that are all equal. No Hamiltonian system with the same number of degrees of freedom (pairs of position and momentum variables) can possess a greater number of positive Lyapunov characteristic exponents and hence be "more chaotic." The present one-dimensional gas therefore shows "maximum hyperchaos."[31] It is thus as chaotic as a gas of billiard balls in three dimensions can ever be (although, paradoxically, the mathematical proof that a 3-D gas is as chaotic as the present one is still wanting).

A reason why the present 1-D gas is so unexpectedly "powerful" can be seen to lie in the fact that it possesses a — from a mechanical point of view — somewhat awkward-to-implement property. In Eq. (1), it has been assumed implicitly that the interactions between the particles are confined to "horizontal vs. vertical" collisions. That is, the particles in the curved part of the tube in Fig. 1 were assumed to be noninteracting. This can be guaranteed mechanically by assuming that each of the two-dimensional loaf-shaped disks runs in its own one-dimensional tracks (with a twist provided before the junction in Fig. 1 is reached so that the particles can interact there). On the other hand, no such "mechanical excuse" is really needed in order to render a Hamiltonian like that of Eq. (1) legitimate mathematically.

Now that we have n particles, we can add "colors" to them, as well as color-changing "collision rules."[31] In this way a 1-D MDS system is obtained which possesses "reaction-kinetic competence." In it, chemical dissipative structures — for example, of the limit cycle type — can now be implemented. The one-dimensionality guarantees that low concentrations do not generate fluctuation-increasing "dilution effects in space" (as is almost unavoidable in two and higher space dimensions).

For example, the kinetics for the simplest reaction scheme producing a limit cycle oscillation[31] may be used. By changing one or two of the parameters, it will be possible to modify the limit cycle in such a way that the system, after performing a single-sweep oscillation, comes to a standstill — until it is very slightly disturbed. The system then becomes a so-called "excitable system." It would be an analog of Winfree's modification of the Belousov–Zhabotinsky oscillator (the no-longer-spontaneously-oscillatory "Z reagent"[45]). The latter constitutes a "fluid neuron."[25]

The obtained excitable system would be a first explicit observer. It would represent the first step on the way to a full-fledged "explicit brain." To obtain the latter, not 1 or 10 or 100 such explicit excitable systems will have to be combined (as presently feasible), but maybe 10 billion. Nevertheless, the above fluid neuron will be the first representative of the whole class. Some of its properties remain valid even for the most complex members of the class. Such properties are of the greatest interest in the present context. A closer look is therefore appropriate.

8.4   An Explicit Measurement Situation

The excitable system (explicit observer) can be combined with a second explicit dissipative structure which acts like an "amplifier." Think of a device analogous to the power-brake in a car. The amplifier, in turn, may be equipped with very light particles at its input end and, moreover, be cooled in its first stage ("preamplifier"). Intermittently, a third explicit dissipative structure could be brought into play to do the cooling. The resulting "measuring chain" would then translate the fine motions of an "object particle" into stronger, more or less proportional but slightly delayed, motions of an output particle (a pressure-exerting piston of larger mass). The latter would directly impinge on the observer as a mechanical "pointer." In this way an explicit measurement situation would be obtained.

When one is simulating this whole system in a computer (which has yet to be done), nothing surprising will happen at first sight. When the pointer pushes strongly enough, macroscopically speaking, the observer will be excited; that is all.

The properties just described are exactly those that one would naively expect from any classical microscopic description of a macroscopic observer who uses a macroscopic measuring device to observe a microscopic particle. Such a classical universe represents progress since it reproduces most of the relevant features of our own world — including the partition between macroscopic intelligent agents called observers on the one hand and macroscopic machinery (including computers) usable as measuring equipment, for example, on the other. A micro level particle could then be observed by the agents with the aid of this advanced "microscope."

In spite of its charm such a classical universe is usually considered useless for understanding our own world for the latter, as is well known, is nondeterministic on the micro level. Furthermore, even if there were something to be learned about the quantum indeterminacy in this fashion, there would still be the nonlocal quantum correlations left which, as is also well known, cannot be reproduced in any standard deterministic classical model according to Bell's theorem.[2]

Unexpectedly, while all of this is true, it is also subtly false. To be more precise, it is correct on the "exophysical level" (considered so far) but false on the "endophysical" one. Let us now turn to the latter.

8.5   Endophysics of the Explicit Observer

The fundamental distinction between exo- and endophysics (not to be confused with "end-o'-physics") is fairly new. See Chap. 6. The distinction between a "physics from without" and a "physics from within"[29,30] is analogous to the distinction introduced by Gödel[12] between generally valid theorems on the one hand and theorems accessible from the inside of the axiomatic system in question on the other. While Gödel's theorem indeed forms a special case, the question focused on in physics is more restricted: How does the explicit mathematical universe in question appear when it is viewed from the inside? More specifically, there are now two questions:

(a) Are there external features left over that are not accessible from the inside? (This is Gödel's question.)

(b) Are there internally valid features that nevertheless are absent when the same system is viewed from the outside?

The latter question was first posed by Maxwell,[17] who showed that to an outside observer (he said "being" or "demon" for lack of a better term), the second law will not exist. The disordered heat motion of a gas at equilibrium can be retransformed into macroscopic mechanical energy (whereby two basic ways are open; see Chap. 6). At the same time, physics must appear "distorted" from the inside. More recently, Finkelstein[8,9] in a discrete context questioned the external validity of both relativity and quantum mechanics, so did Fredkin and Toffoli.[10] The same hypothesis was also introduced in a continuous context.[29,39,9a,40a]

How does the present explicit universe appear to the internal observer? To see this, it is necessary to give the dynamics already considered a "second look." The first to do so in the context of a classical gas at equilibrium was Gibbs.[11] Unexpectedly, there is a symmetry principle at work. The latter drastically changes the behavior even after one had already thought there was nothing left to be learned about the system. Gibbs' symmetry is implicit in Eq. (1) above. The different particles were assumed to be equal in all physical and mathematical properties (except for position and momentum) when belonging to the same class, including color.

Gibbs showed that under this condition, phase space volume has to be divided by $N!$ in the simplest case (when there are N equal particles of a single class present), and by $N_1! \cdot N_2! \ldots$ in general (when there is more than one such class).[11]

The result was later found to be mathematically sound, when Weyl[42] rediscovered it independently; cf. Refs. 34 and 34a. Qualitatively it all goes back to Leibniz[15] and Spinoza.[12,34] The essence of the principle goes as follows. Particle indistinguishability (in the sense of mathematical equality) is an admissible hypothesis in physics since particle-like solutions of fields (so-called solitons; cf. Ref. 46) possess the same property. It implies "trajectorial multiuniqueness" in phase space,[31] a property apparently first glimpsed by Leinaas and Myrheim.[15a] This is a new mathematical feature of certain systems of ordinary differential equations, including most Hamiltonian systems. The new axiom has new mathematical implications.[34]

In short, the system in question becomes confined to a subcell of its phase space — one of $N!$ identical cells that together make up the full phase space. This group-theoretic reduction[31] is not a matter of mere descriptive economy. It has measurable consequences, as Sackur[37] showed. The fact that the phase space volume of a classical gas goes down by the factor $N_1!$ in the presence of N_1 equal particles is reflected in the measurable equilibrium entropy of a hot gas;[37] cf. Ref. 5a.

Similar findings hold good for the subsystem of interest in the present case, i.e. the explicit observer. This is perhaps best seen with the aid of an example. Assume that the observer consists of just two identical particles (both red), with a third particle (blue) making up the "rest of the universe."

This three-particle system is exophysically described by Eq. (1) with $n = 3$. Figure 3 above already showed the position space — in the absence of the Gibbs symmetry. Indistinguishability between two of the three particles implies trajectorial biuniqueness (since $N_1 = 2$ yields $N_1! = 2$). Trajectorial biuniqueness then has the consequence that phase space volume becomes halved. The result is presented in Fig. 4.

At the same time a new, "endophysically correct" Hamiltonian replaces Eq. (1). It reads

$$H' = H + \frac{\varepsilon}{x_2 - x_1} . \tag{2}$$

H is the Hamiltonian of Eq. (1), without the exchange symmetry, and ε approaches zero.[31]

The blue (third) particle no longer interacts with equal probability with both red particles — as it would in the absence of the symmetry — but only with the first red one when it is up itself (and only with the second red particle when it is down itself) in the tube of Fig. 1. Note that according to Eq. (2), the first red particle (described by x_1) is now confined to the lower values of x while the second preferentially occupies the higher values. This is because the two equal particles cannot pass by each other any more, endophysically, since they necessarily exchange their identities whenever they pass by each other in a naive exo sense.[31,34]

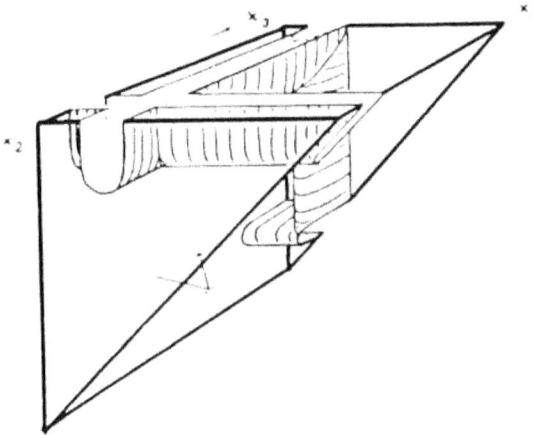

Fig. 4. Reduced configuration space for the system of Fig. 3. The cutting-in-half is caused by the presence of an exchange symmetry valid between two of the three particles (cf. text). A trajectory is indicated as in Fig. 3.

This can also be seen geometrically. In Fig. 4, only one of the two vertical bars of Fig. 3 is left (the one at the back of the cell). This bar corresponds to the interaction between the blue and the first red particle (x_3 and x_1), as Fig. 3 shows. The symmetric interaction between the same two particles (when x_1 is down and x_3 is up), corresponding to the vertical bar seen in the foreground of Fig. 3, now belongs to the other cell.

In general, when more than two red particles ($N_1 > 2$) make up the "observer" and the blue particle still forms the "rest of the universe," the new Hamiltonian contains $N_1 - 1$ terms of the type present in Eq. (2), namely

$$H' = H + \varepsilon \left(\frac{1}{x_N - x_{N-1}} + \frac{1}{x_{N-1} - x_{N-2}} + \cdots + \frac{1}{x_2 - x_1} \right). \qquad (3)$$

where N has been written for N_1 for simplicity.[31]

What does the resulting reduction of phase space volume imply for the observer?

8.6   Implications of the Reduced Phase Space

One would expect the new dynamics to be, while quantitatively different from the old one, at least qualitatively the same as before. Not even this is the case, however. While the old dynamics was maximally chaotic, the new dynamics is essentially nonchaotic, namely, quasiperiodic.[31,33] The reason lies in the fact that equal particles can no longer pass by each other but seemingly repel each other elastically. Most "inner" particles in the ring of Fig. 1 therefore no longer reach the point where the action (chaos generation) is. For only in the region of large or small (respectively) x values is trajectorial divergence generated.

The new Hamiltonian, Eq. (3), clearly generates different trajectories than the original one, Eq. (1). Only locally is there still a coincidence with one of the many original (multiunique) trajectories.[31]

Armed with this main result, we can ask for its implications. Sixteen were previously described (Chap. 6, Secs. 6.8–6.11). The main implication was that any external object that is directly coupled to the observer is subject to an effective diffusion coefficient, of size

$$D = \frac{ET}{M} \tag{4}$$

Here M is the object's mass, T the mean cell passage time inside the observer, and E the mean kinetic energy per particle in the observer. A directly coupled external object therefore performs a Brownian motion, determined by D, relative to the observer. This is not too surprising since, apart from the usual thermal noise energy E, only T enters, a characteristic period arising within the observer. The noise energy E is effectively reapplied to the external object after every unit time interval T. This leads directly to Eq. (4), as we saw in Chap. 6.

Unexpectedly, the same result, Eq. (4), holds good for "indirectly coupled" objects. Indirectly coupled objects are, for example, objects that are observed by the observer as the latter makes use of a measuring device. This latter result, if correct, is surprising. The reason is that Eq. (4), with $h/4\pi$ standing for ET, is well known in quantum mechanics — where it is called Nelson's postulate. Nelson[18] postulated this equation as an axiom and showed that this single assumption suffices for obtaining the Schrödinger equation and hence most of quantum mechanics.[38] This would mean that Nelson's stochastic mechanics (and therefore quantum mechanics) is an implication of the present classical universe.

The question whether Eq. (4), with fixed ET, is indeed an endophysical implication of the universe of Eq. (1) is, therefore, the main question (No. 1) to be considered in the following. A second question immediately follows (No. 2): If stochastic mechanics follows from classical first principles, how about the quantum nonlocality? (Nelson[19] so far only showed a compatibility-in-principle.) Thereafter, the following two questions would come in as a kind of "coda." First, (No. 3), how can we reconcile the two facts that, on the one hand, the pointer suffers no appreciable uncertainty according to Eq. (4) by virtue of its large mass while, on the other, the low mass object whose motion is proportionally represented by the pointer suffers a strong uncertainty by the same equation? Finally (No. 4): How does the fact that ET is not a universal constant, but a constant only for the observer, make itself felt to the observer?

8.7   A New Principle

The main result underlying the above approach was "trajectorial $N!$-uniqueness" in the presence of particle indistinguishability. This result, it turns out, is closely related to an even more general principle: "trajectorial biuniqueness" in *all* Hamiltonian systems of the Newtonian type. That is, the number of trajectories in phase space doubles once more. This second principle, to be demonstrated next, will then provide answers to the four questions posed in the preceding section.

Recall the role of the "quasiperiod" T. The observer is effectively subjected to an oscillation with this mean half-period (Chap. 6). But why should that fact greatly affect his observations? The reason was that every typical motion inside the observer, and every external motion, reverse their mutual orientation after every unit time interval T. Therefore, so the reasoning went, the relative relation is the same as if the observer had not changed his own orientation in time, but rather the external particle (and the whole measuring chain) had done so.

This reasoning appears flawless. If time does not flow (as is the case in science[5,29]), the two situations must be equivalent. On the other hand, the question of time reversal invariance, which is thereby elevated to unexpected new prominence, certainly deserves to be readdressed in its own right before the final verdict.

The new group-theoretic instrument (analysis of multiunique traject-ories[31]) provides a solution to this new question as well. In fact, the main point was already recognized by Wigner in a quantum context.[44] Only later did the evidence accumulate that classical systems require an analogous treatment.[16,5c]

Let us adapt Wigner's result to the present context. Hamiltonian systems in general possess *two* trajectories instead of a single one (a "black" and a "white" one for short). The black trajectory is the one usually considered. As an example, let us take an ordinary single-particle Hamiltonian H of the Newtonian type. For the given Hamiltonian H, quadratic in p, Hamilton's recipe (written in brackets on the right hand side below) yields the following two differential equations of motion:

$$\frac{dx}{dt} = p \left(= +\frac{dH}{dp} \right) ,$$

$$\frac{dp}{dt} = f \left(= -\frac{dH}{dx} \right) \tag{5}$$

Here f is the Newtonian force. (For example, in the case of the harmonic oscillator, $H = \frac{1}{2}x^2 + \frac{1}{2}p^2$, we have $f = -x$.) The corresponding initial conditions are x_0 (the given initial position) and $p_0 = mv_0$ [v_0 is the initial rate of change of x, i.e. $(dx/dt)_0$, and $m = 1$]. This is the usual way of writing down the equations of motion. It yields the black trajectory.

However, there is a second, equally correct trajectory implied by H. This white trajectory possesses the following equations of motion:

$$\frac{dx'}{dt} = -p' \left(= -\frac{dH}{dp'} \right) ,$$

$$\frac{dp'}{dt} = -f \left(= +\frac{dH}{dx'} \right) . \tag{6}$$

It has the initial conditions $x'_0 = x_0$ and $p'_0 = -p_0$.

The bijective transformation rule, valid between Eqs. (5) and (6), reads $x' = x$ and $p' = -p$. At first sight this rule appears innocuous. The second description can be uniquely related to the first, being its reflection at the $p = 0$ axis in the present case (in general, it is reflected at the origin of momentum space). The world seemingly can be fixed again since the two descriptions are identical under an automorphism (a reflection). This impression is misleading, however, since the two descriptions are nonequivalent under time reversal.

Equation (6) at first sight appears to be equivalent to Eq. (5) since it transforms into it under time reversal. That is, inverting time ($t' = -t$) in Eq. (6), one obtains

$$\frac{dx'}{dt'} = p' \left(= +\frac{dH}{dp'} \right) ,$$

$$\frac{dp'}{dt'} = f \left(= -\frac{dH}{dx'} \right) ,$$

(6')

with initial conditions $x'_0 = x_0$ and $p'_0 = -p_0$ as before, since only the direction of flow but not the initial condition is changed under mathematical time reversal, $t' = -t$.

As expected, Eq. (6') is identical to Eq. (5) if both x and x' and p and p' are identified and so are t and t'. However, there is a remaining difference: the initial conditions are different! They were x_0 and p_0 in Eq. (5) and are x_0 and $-p_0$ in Eq. (6'). The former pair determines the black trajectory in both directions of time; the latter pair determines the white trajectory in both directions of time. Thus, there indeed exist two trajectories in general.

Wigner[43] said it all in one sentence: "Physical time reversal is motion reversal." He could have said: "... is momentum reversal." He could even have said: "... is reversal of initial momentums." That is, if *only* the signs of all initial velocities are changed, physical time reversal has been accomplished. This is paradoxical. Time need not be inverted at all to obtain physical time reversal. One also sees that the relabeling of particle identities, as a boundary of the phase space granule is reached, dramatically increases the chances for a time reversal (a reversal of all momenta) to be possible and to — indeed — occur spontaneously.

To come back to our context, a switch from the black to the white trajectory suffices for time reversal. Or, to back-pedal still further for a moment, Hamiltonian systems in general possess two trajectories in their phase spaces. Both are mirror images obtained from each other through reflection at the point $p_i = 0$ (all momentums equal zero), as mentioned. Nevertheless neither can be reduced to the other. Thus, if one wants to introduce velocities at all (which was deliberately shunned by Newton, who stuck to second order differential equations, as is well known), nonuniqueness can be avoided only if the indistinguishability of the two directions of time is taken into explicit consideration — as a symmetry. This much is in principle well known; cf. Ref. 5c.

What is new, perhaps, is that this is nothing but another instance of trajectorial biuniqueness. This fact has consequences. Specifically, we have two: (A) the exophysically correct phase space collapses down to half its former volume, once more; (B) in the obtained "reduced phase space," trajectorial uniqueness is re-established, once more. That is, the new unique trajectory differs from either predecessor in that it consists, alternatingly, of "black" and "white" pieces connected by "jump segments" of zero duration,[32] once more.

This, however, is exactly the lacking piece of information that is needed to answer the first (No. 1) question of Sec. 8.6: the "black" and "white" time slices which had been presupposed in the derivation of Eq. (4) do indeed exist.

Let us now turn to the second (No. 2) question of Sec. 8.6.

8.8   Nonlocality

It goes without saying but nevertheless it is important to realize that the Brownian motion described by Eq. (4) is not something that exists objectively (exophysically) in the universe assumed. Rather, it exists by definition only for the observer (endophysically). Nevertheless it represents inescapable reality, to the observer. It is an objective property of the "interface" between the observer and the rest of the universe. In this sense it is universal – observer-universal.

The very nonobjective validity of the observer-specific action (ET) is the reason why its implications are not covered by Bell's[2] theorem, which states that deterministic hidden variables of the classical local type can never reproduce the quantum nonlocality. Thus, *if* the present theory turned out to imply the quantum nonlocality, this "endo" counterexample would not violate the original (exo) theory.

To see whether nonlocal correlations can indeed arise in a purely endophysical way, let us consider two exophysically correlated objects (having zero combined momentum, for example) in the model universe. Since the observer generates all perturbations through his own passing through black and white time slices (trajectorial segments) in a particular fashion, the two objects will be perturbed identically. For they both reflect the same measurable state of the same composite original particle at the moment of splitting in two.

Therefore a second measurement, made with a different aim on the other particle (measuring momentum rather than position, for example), will necessarily show the same outcome as if the first measurement had been made on that particle, too. The fact that it was made on the other makes no difference since that measurement had really been a measurement on the properties of the disintegrating combined particle and hence on the properties of both. A measurement of the same type, however, is at the root of Bell's own example (although transposed to the context of angular momentum rather than ordinary momentum). An analog of the unilaterally valid ($\cos^2$) law[38] therefore also applies in the bilateral case. An inequality analogous to that based on Malus' law found by Bell[2] can thus be predicted to be valid in the present case.

This means that apart from stochastic mechanics, also the quantum nonlocality appears to be an implication of the present deterministic local hidden variables universe.

8.9 Conclusions

A microscopically exact view of a conceptually transparent universe has been proposed. The universe was chaotic, classical and local. The implications nevertheless were nonchaotic, nonclassical and nonlocal. This is surprising enough to warrant a critical second look. What is the weakest point?

Probably, it lies not so much with what has been shown as with what has not been shown, namely: How can a nonperturbed massive pointer mediate the perturbed behavior of a light object particle when the two are actually strongly correlated? This third question (No. 3) of Sec. 8.6 is much harder to answer than were its two antecedents, stochasticity and nonlocality.

There is hope that the question can eventually be answered using the "x, t plot" technique (all particle positions vs. time). The latter plot contains all information and indeed allowed the first spotting of the time slices.[32] The optimal way of compressing and processing this information so that it makes intuitive sense again has yet to be found, however. Therefore only a qualitative argument can be offered at the time being, in the form of a hypothesis.

The contradiction can be resolved only if the pointer particle is not registered exo-objectively while it is a part of the measuring chain. Endophysically, the measuring chain turns black and white in synchrony with the observer; exophysically, it does not. The endo distortion therefore is absent on the exo level. Nevertheless the pointer particle must appear like any other macro particle when looked at as an object in its own right. Niels Bohr referred to a related problem when he claimed that a blind man with a cane can either feel "with" the cane or feel "the" cane.[6] Accomplishing the two feats simultaneously is impossible. This, however, is precisely the task required here.

The following possibility comes to mind. The pointer could appear "frozen out" in order to represent the momentarily valid perturbed behavior of the object. At the next moment, with a different sequence of black and white time slices in the observer in charge, the object has to appear perturbed differently to the observer. The pointer will in that case have to appear "frozen out" differently, and so forth. In other words, the whole macroscopic world (including the pointer) and its past must appear consistent in supporting a certain perturbed state of the object at every moment, but it must do so differently at every moment. This implies that the change involved is not accessible to the observer. Thus a "masking principle," shielding the observer from inconsistency-generating information, is invoked.

This "desperate" proposal may turn out to be correct (verifiable). Unexpectedly the proposal at the same time represents a possible solution to the measurement problem of quantum mechanics. It is closely related to two recent interpretations put forward by Bell[3] and Deutsch[5b] (cf. also Ref. 21) in the context of Everett's[7] so-called many-worlds interpretation of quantum mechanics: At every moment, a different Everett world is in charge without this fact being accessible to the observer.[3] "Nowness" thereby becomes codimension-2 (there is a second free parameter) rather than merely codimension-1 (a point on the time axis). Thus Everett's version of quantum mechanics is effectively implied by the above proposal.

Simultaneously the remaining question (No. 4) of Sec. 8.6 is taken care of by the masking principle. For if every single individual momentary world is internally consistent, the observer not only does not register mutually deviating individual outcomes, valid in different worlds, but also does not register mutually deviating mean values obtained from many past measurements, valid in different worlds. Hence a constant "universal value" is always seemingly represented.

Thus the three major riddles of quantum mechanics (stochasticity, nonlocality, state reduction) have all resurfaced in the artificial universe, either as an implication (the first two) or as a proposed implication (the third). A new, fourth problem (universality) arose but turned out to be a special case of the third. All of the three implications found appear to be typical, not only for the present reversible universe, but also for "most" reversible universes.

There may be further major implications of reversible universes. Let us, however, turn to a more general question. Suppose an artificial universe like the one described above were actually built some day. Suppose further that developments in computer technology had made it possible to implement many excitable systems (formal neurons) simultaneously so that an arbitrarily complicated reversible artificial brain (or several) could be built. Would the inhabitants be able to find out? A similar question was first proposed in a dissipative context by Putnam.[22] He convincingly argued that the answer must be in the negative. In the present reversible context, a positive answer appears to be possible.

Let us therefore finish by describing an experiment ("meta-experiment") which the inhabitants would be able to perform after having embarked on an endophysical program of their own. The experiment is called the "fever test."

The observer first calculates his own ET on the basis of available information (mass-density of the smallest particles that make him up, and body temperature). Then he compares this value with the universal ET (quantum of action) from the books. If the two coincide to one digit, he tries to improve on the result by measuring and calculating more accurately until the coincidence becomes very good (this would obviously be possible in the model universe). Now begins the second phase. He takes a fever pill and redoes the calculation with the new data collected while having a fever. He predictably finds a deviation in the fourth digit between his own calculated ET and that of his friends, who do not show a fever. Then he again makes a comparison with the books. The experiment is rated a success if one of the calculated numbers consistently turns out to be privileged — his own.

In the model universe, this activity would be compounded by a "sociological" problem. While the experimenter could easily perform the experiment as described, he would have difficulty convincing others of its seriosity. Even the first step — the single-digit coincidence assumed for starters — would be considered an unpublishable observation by his colleagues. In their nonrelativistic (up to that moment) world, there would never have been any reason (like a superluminal nonlocality) to question the objective reality of the quantum laws. The endophysical insight that interfaces are something objective too would come less easy to them. Moreover, our own world differs from theirs in an additional respect. So many important forces (radiation, gravity, strong force) were omitted in setting up the artificial universe that the latter does not quite qualify as a model. This makes the exotic fever test appear acceptable to us as a tale about the lower level.

To conclude, reversible worlds are different. Owing to the Wigner symmetry, external causality vacillates for every subsystem. The implications regarding observation are not yet fully understood. They are bound to be far-reaching. So far, only some fairly "tame" consequences which resemble quantum mechanics have been pinpointed. Nevertheless Parmenides' and Kant's assertion that reality is different from the way it appears can already be confirmed — in a toy setting. The science of artificial life can do more than open up new worlds for us: it may redefine our own.

Acknowledgments

This chapter is based on an invited paper presented under the title "Explicit Observers in an Invertible Artificial World" at the first "Artificial Life Workshop," held at the Center for Nonlinear Studies, Los Alamos, September

21–25, 1987. I thank Peter Plath, Chris Langton and Gottfried Mayer-Kress for stimulation and Art Winfree, Anatol Zhabotinsky, Norman Packard, Doyne Farmer, Jim Crutchfield, Jeffrey Tennyson, Igor Gumowski, Christian Mira, Dietrich Hoffmann and Hans Primas for discussions.

References

1. B. J. Alder and T. E. Wainwright, "Phase transitions for a hard-sphere system," *J. Chem. Phys.* **27**, 1208–1209 (1957).

2. J. S. Bell, "On the Einstein–Podolsky–Rosen paradox," *Physics* **1**, 195–200 (1964).

3. J. S. Bell, "Quantum mechanics for cosmologists," in *Quantum Gravity*, Vol. 2, eds. C. J. Isham, R. Penrose and D. W. Sciama, (Oxford University Press, 1981), pp. 611–637.

4. B. J. Berne, *Statistical Mechanics, Part B: Time-Dependent Processes* (Plenum, New York, 1977), pp. 41–63.

4a. L. Bunimovich and Ya. G. Sinai, *Commun. Math. Phys.* **78**, 247 (1980).

5. P. C. W. Davies, *The Physics of Time Asymmetry* (Surrey University Press, 1974).

5a. K. G. Denbigh and J. S. Denbigh, *Entropy in Relation to Incomplete Knowledge* (Cambridge University Press, Cambridge, 1985).

5b. D. Deutsch, "The connection between Everett's interpretation and experiment," in *Quantum Concepts in Space and Time*, eds. R. Penrose and C. J. Isham (Clarendon, 1986), pp. 215–225.

5c. R. L. Devaney, "Reversible endomorphisms and flows," *Trans. Amer. Math. Soc.* **218**, 89–96 (1976).

5d. H. H. Diebner and O. E. Rössler, "A deterministic entropy based on the momentary phase space volume," *Z. Naturforsch.*, submitted.

5e. The following paper contains an example in which 1000 point-shaped billiard disks are used to implement a marginally stable oscillatory reaction system: H. H. Diebner and O. E. Rössler, "Deterministic continuous MDS of a chemical oscillator," *Z. Naturforsch.* **50a**, 1139–1140 (1995).

6. P. A. M. Dirac, "The versatility of Niels Bohr," in *Niels Bohr*, ed. S. Rozental, (Elsevier, Amsterdam, 1967), pp. 306–309.

7. H. Everett, "'Relative-state formulation' of quantum mechanics," *Rev. Mod. Phys.* **29**, 454 (1957).

8. D. Finkelstein, "Holistic methods in quantum logic," in *Quantum Theory and the Structures of Time and Space*, Vol. 3, eds. L. Castell and C. F. von Weizsäcker (Carl Hanser, Munich, 1979), pp. 37–59.

9. D. Finkelstein and S. R. Finkelstein, "Computer interactivity simulates quantum complementarity," *Int. J. Theor. Phys.* **22**, 753–779 (1983).

9a. J. Ford, "Directions in classical chaos," in *Directions in Chaos*, ed. Hao Bai-Lin (World Scientific, Singapore, 1987), pp. 1–16.

10. E. Fredkin and T. Toffoli, "Conservative logic," *Int. J. Theor. Phys.* **21**, 219–253 (1982).

11. J. W. Gibbs, *Elementary Principles in Statistical Mechanics* (Yale University Press, New Haven, 1902), Chap. 15.

12. K. Gödel, *On Formally Undecidable Theorems* (Basic Books, New York, 1962) (originally published in 1931). Cf. also: P. Weibel and E. Köhler, "Gödel's undecidability proof: contour lines in the history of ideas concerning an important mathematical theorem" (in German), in *Gödel-Satz, Möbius-Schleife, Computer-Ich*, ed. F. Kreuzer (Franz Deuticke Verlagsgesellschaft, Vienna, 1986), pp. 77–101.

13. M. Heinrichs and F. W. Schneider, "Molecular dynamics calculations of a second-order kinetic phase transition in an open system (CSTR)," *Ber. Bunsenges. Phys. Chem.* **87**, 1195–1201 (1983).

14. J. L. Hudson and O. E. Rössler, "Chaos and complex oscillations in stirred chemical reactors," in *Dynamics of Nonlinear Systems*, ed. V. Hlavacek (Gordon and Breach, New York, 1986), pp. 193–219.

15. G. W. Leibniz, *The Leibniz Clark Correspondence*, ed. H. G. Alexander (Manchester University Press, Barnes and Noble, Manchester, 1956), pp. 26, 38, 63.

15a. M. Leinaas and J. Myrheim, "On the theory of identical particles," *Il Nuovo Cimento* **B37**, 1–23 (1977).

15b. E. N. Lorenz, "The problem of predicting the climate from the governing equations," *Tellus* **16**, 1–10 (1966).

16. G. Lüders, "On the motion reversal in quantized field theories" (in German), *Z. Phys.* **133**, 325–339 (1952).

17. J. C. Maxwell, *Theory of Heat* (Appleton, New York, 1872), p. 309.

18. E. Nelson, "Derivation of the Schrödinger equation from Newtonian mechanics," *Phys. Rev.* **150**, 1079–1085 (1966).

19. E. Nelson, "The locality problem in stochastic mechanics," *Ann. N. Y. Acad. Sci.* **480**, 533–538 (1987).

20. G. Nicolis and I. Prigogine, *The Investigation of the Complex* (in German) (Piper, Munich, 1987).

21. D. Page and W. K. Wootters, "Evolution without evolution — physics described by stationary variables," *Phys. Rev.* **D27**, 2885–2892 (1983).

21a. I. Prigogine and I. Stengers, *Time, Chaos and the Quantum* (German edn. *Das Paradox der Zeit — Zeit, Chaos und Quanten*) (Piper, Munich, 1993).

22. H. Putnam, *Reason, Truth and History* (Cambridge University Press, Cambridge, 1981), Chap. 1.

23. O. E. Rössler, "A system-theoretic model of biogenesis" (in German), *Z. Naturforsch.* **26b**, 741–746 (1971).

24. O. E. Rössler, "Design for autonomous chemical growth under different environmental constraints," *Prog. Theor. Biol.* **2**, 167–211 (1972).

25. O. E. Rössler, "A synthetic approach to exotic kinetics (with examples)," *Lect. Not. Biomath.* **4**, 546–581 (1974).

26. O. E. Rössler, "Adequate locomotion strategies for an abstract organism in an abstract environment: a relational approach to brain function," *Lect. Not. Biomath.* **4**, 342–369 (1974).

27. O. E. Rössler, "Chemical automata in homogeneous and reaction–diffusion kinetics," *Lect. Not. Biomath.* **4**, 399–418 (1974).

28. O. E. Rössler, "Chaotic behavior in simple reaction systems," *Z. Naturforsch.* **31a**, 259–264 (1976).

29. O. E. Rössler, "Chaos and chemistry," in *Nonlinear Phenomena in Chemical Dynamics*, eds. C. Vidal and A. Pacault (Springer-Verlag, New York, Heidelberg, 1981), pp. 79–87.

30. O. E. Rössler, "Macroscopic behavior in a simple chaotic Hamiltonian system," in *Dynamical Systems and Chaos*, ed. L. Garrido, *Lect. Not. Phys.* **179**, 67–77 (1983).

30a. O. E. Rössler, "Classical quantization — two possible approaches," in *Chaotic Behavior in Quantum Systems: Theory and Applications*, ed. G. Casati (Plenum, New York, 1985), pp. 345–351.

31. O. E. Rössler, "A chaotic 1-D gas — some implications," in *The Physics of Phase Space*, eds. Y. S. Kim and W. W. Zachary, *Lect. Not. Phys.* **278**, 9–11 (1987).

32. O. E. Rössler, "Endophysics," in *Real Brains, Artificial Minds*, eds. J. L. Casti and A. Karlqvist (North-Holland, New York, Amsterdam, 1987), pp. 25-46. Cf. Chap. 6 of this book.

33. O. E. Rössler, "Explicit dissipative structures," *Found. Phys.* **17**, 679–688 (1987).

34. O. E. Rössler, "Symmetry-induced disappearance of reality: the Leibniz effect," in *Art and the New Biology: Biological Forms and Patterns*, ed. P. Erdi, *Leonardo* **22**, 55–59 (1989).

34a. O. E. Rössler, "Jumping identities of particles," *Symmetry: Culture and Science* **7**, 307–319 (1996).

35. O. E. Rössler and C. Kahlert, "Winfree meandering in a 2-dimensional 2-variable excitable medium," *Z. Naturforsch.* **34a**, 565–570 (1979).

36. O. E. Rössler and M. Hoffmann, "Quasiperiodization in classical hyperchaos," *J. Comp. Chem.* **8**, 510–515 (1987).

37. O. Sackur, "Applying the kinetic theory of gases to chemical problems" (in German), *Ann. der Phys.* **36**, 958–980 (1911).

38. W. R. Schneider, "Stochastic mechanics," in *Quantum Mechanics Today*, Lecture Notes of 11th Gwatt Workshop, October 15–17, 1987, eds. D. Baeriswyl, M. Droz, C. P. Enz and A. Malaspinas (BBC, CH-5405, Baden, 1987), pp. 234–251.

39. R. Shaw, "Strange attractors, chaotic behaviour and information flow," *Z. Naturforsch.* **36a**, 80–112 (1981).

40. Ya. G. Sinai, "Dynamical systems with elastic reflections," *Russian Math. Surveys* **25**, 137–190 (1970).

40a. K. Tomita, "Conjugate pair of representations in chaos and quantum mechanics," *Found. Phys.* **17**, 699–711 (1987); reprinted in *A Chaotic Hierarchy*, eds G. Baier and M. Klein (World Scientific, Singapore, 1991), pp. 353–363.

41. R. Wegscheider, "On simultaneous equilibria and the relations between the thermodynamics and the reaction kinetics of homogeneous systems" (in German), *Z. Phys. Chem.* **39**, 257–303 (1902).

42. H. Weyl, *Philosophy of Mathematics and Science* (Princeton University Press, Princeton, 1949), Chap. 22e, App. B.3.

43. E. P. Wigner, "Relativistic invariance and quantum phenomena," *Rev. Mod. Phys.* **29**, 255–268 (1957).

44. E. P. Wigner, *Group Theory and Its Application to the Quantum Mechanics of Atomic Spectra* (Academic, New York, 1959, pp. 325–348; pp. 333, 364; originally published in 1932). Cf. also: M. G. Doncel, "From magnet to time reversal," in *Symmetry in Physics 1600–1980*, eds. M. G. Doncel, A. Hermann, L. Michel and A. Pais (Universitad Autonoma de Barcelona Bellaterra Press, Barcelona, 1987), pp. 409–429.

44a. K. D. Willamowski and O. E. Rossler, "Irregular oscillations in a realistic abstract quadratic mass-action system," *Z. Naturfosch.* **35a**, 317–318 (1980).

45. A. T. Winfree, *The Geometry of Biological Time* (Springer-Verlag, New York, Heidelberg, 1980), p. 240.

46. I thank Richard Bagley for pointing out to me in 1987 that solitons (particle-like solutions which arise in Hamiltonian partial differential equations) indeed come in discrete classes in two dimensions — so that indistinguishable "particles" do indeed exist mathematically.

9
The Endo Approach

Coauthor: Jonas Rössler

Abstract

Like a virtual reality lacking an escape button, the world can be studied only from within. Nevertheless, lower level worlds can be built in which the "interface" between an explicit observer and the rest of the universe can be scrutinized. The interface depends sensitively on microscopic assignment conditions ("which particle belongs to the observer, which to the rest"). Furthermore, the successive distorted interfaces (observer-relative worlds) are neither constant nor intertransformable. They form an "objective reality of the subjective type." An example is an advanced classical molecular dynamics simulation of the future. Here an excitable subsystem (fluid neuron) is the "observer," a cooled fluidic amplifier the "measuring chain" and a microparticle the "object." Unexpectedly, knowledge of the macrostate of the observer and the "pointer" of the measuring chain is insufficient for predicting the reality on the interface. This can be seen best from a simplified case. The observer is a collection of pendulums of equal periods (T) and equal phases, but with the signs of the phases randomly different. All motions in the rest of the universe then acquire a "zigzag" character relative to the observer. The resulting distortion of the world depends on a further parameter, the thermal noise energy of the observer (E). After every micro-time-reversal T, this energy is imposed on the object through the measuring chain, which then acts in the other direction. Bohr's example of the blind man's cane is re-encountered: it is impossible to simultaneously get a tactile sensation of an object by means of a cane and feel the end of the cane as it is held in one's hand. Similarly, the high-mass pointer on the measuring apparatus can only mediate the distorted behavior of the micro-object in a "frozen" fashion. Both the zigzag and its frozen representation are only a mirage. In the next interface valid at the next moment in time, both are different. However, the difference is nowhere represented — so it does not exist, in the mirage. Nevertheless, an empirical "unmasking" is possible within the model universe. Analogous "blind-sight experiments" can be performed on our own level. Therefore it will be possible to find out whether the real world too is only an interface reality — a mirage.

Acknowledgments

The first author would like to thank Michael Conrad and John Casti for their kindness also in the name of the second author.

10
Boscovich Covariance

Summary

A quarter of a millennium ago, mathematical physicist Roger Joseph Boscovich proposed a general physical equivalence principle: (1) whether the observer is in a state of motion relative to the world, and vice versa, are equivalent situations (classical relativity); (2) whether the observer is subject to an internal motion (of the periodic type, say) and whether instead the world is subject to an opposite motion (so that again the two motions cancel out) are also equivalent situations. The latter principle amounts to a generalization of relativity in an unexpected direction. Boscovich illustrated it with the example of a "breathing" world. Apparently, he was motivated to invent his principle of the symmetric total view by the fact that the world appears more solid to us than his own classical solid state theory allows for. He not only was the first to discover a nonclassical feature of the world, he was also the first to explain it with a rational hypothesis. The scope of his principle has yet to be determined — whether all quantum properties can be derived from it, for example. To facilitate the reception of his ideas, a new translation of Boscovich's seminal paper of 1755 has been appended to this chapter.

10.1 Introduction

A kitten which is restrained from exploring the world actively — by being carried around passively in a basket by the exploring locomotion of another kitten — never learns the motion-independent invariant properties of the world.[1] The covariance — that is, the lawfulness present in all the different "covarying" (with displacement) distorted aspects of the world[2] — can only be discovered actively by an intelligent animal. This lends credit to Poincaré's prediction[3] that regardless of what group structure the world may have, there is a mechanism in the brain allowing this structure to become second nature to us after sufficient exploration. Humans can go one step further and discover through armchair reflection a second covariance that is based, not on the position, but on the state of motion. In analogy to the position-specific perspectives in an invariant geometry learned by the kitten, we learn to live with the velocity-specific frames in an invariant hypergeometry, Einstein[4] found out. Surprisingly, the series can be continued. In the following, a third kind of covariance will be described which was first seen by Boscovich.[5]

10.2   Boscovich's Equivalence Principle

Boscovich is known to be one of the forerunners of Einstein (after Cusanus, Galilei, Leibniz and Berkeley) in the context of the discovery of relativistic covariance, as Alexander stressed.[6] However, the fact that Boscovich was unaware of the motion independence of the speed of light diminishes the impact of his insight that the state of motion of an observer relative to his or her world constitutes a primary reality. No similar qualification is necessary when it comes to appreciating his second, related discovery which apparently has gone unnoticed so far.

Boscovich, who is also the inventor of Laplace's demon (Laplace had copied the later-to-be-famous passage verbatim from Boscovich's book[7]), was the first physicist to apply Newton's theory to the microscopic realm. He founded solid state physics and theoretical chemistry by indicating the first potential function for (as one would say today) an atom; see Fig. 1 on page 42 of his book[5] (reproduced by Heilbronner[28]). This Lennard–Jones-like (but more wiggly) potential also appears to be the source of his insight that, contrary to appearances, solid matter (and the world) cannot be rigid but must be more malleable than meets the eye. At any rate, he proposed a new general principle stating that the world must be described relative to an observer.

More specifically, Boscovich claimed that the observer can never observe the world as it is. He can only describe the interface (or "difference") between himself and the world. A first logical deduction from this principle was the finding that a state of external motion of the observer relative to the world is equivalent to a state of motion of the whole world relative to a stationary observer. This insight is at the basis of Einstein's special theory of relativity.[6] Boscovich's seminal paper of 1755, which he appended as a supplement to all editions of his textbook,[5] is to be found at the end of the present chapter in a new translation. The similarity of Boscovich's example of the ship and the shore (Sec. 3 of the appendix) to Einstein's example of two trains standing in a train station with only one pulling out is striking.

Being one of the forerunners of kinetic theory (although the fine motions eluded him), Boscovich was equipped to see a second implication. The same equivalence principle applies also to a state of *internal* motion of the observer relative to the rest of his universe. If, for example, both the observer's internal motions and all motions of his environment suffered a time reversal (a possibility which falls within the scope — if not the letter — of Boscovich's considerations), nothing would change for the observer. For the interface (difference) between the two stays unaffected.

If joint time reversals appear too artificial an assumption, there is a more realistic corollary: if only the observer's internal motions are time-inverted, this is equivalent to no change in the observer but the external world having been time-inverted instead. This means, more generally, that any change inside the observer that can be exactly compensated for by some external change in the environment (so that the net effect on the interface would be zero) is equivalent to that imaginary compensatory change having actually occurred in the environment at the expense of no change having taken place in the observer.

This is an unexpected result. It means that the observer does not see the world as it is. Since the existence of such "in-principle-compensable-for" changes in the observer can never be excluded *a priori*, the observer *never* sees the world objectively. Only a "cut" (or "distortion") dependent on the state of his own internal motions can be accessible to the observer in principle. Moreover, since the interface is the only reality he has, to the observer the world objectively possesses all those features which it only adopts relative to the state of his own internal motions.

Thus Boscovich has discovered a generalization of Einsteinian relativity in a direction not seen by Einstein himself. The situation is like a playback of that created by Einstein covariance. In that earlier case, there suddenly existed a directly inaccessible hyperreality (Minkowski's "absolute world") which determined reality proper. In the newly revived case, it is Boscovich's "true existence modes" which play the same role. However, there is an important difference: Boscovich covariance is distinguished from Einstein covariance in the degree of its accessibility: it is much harder to unmask.

10.3 Opacity

In the case of Boscovich's covariance, even an indirect route to the new hyperreality (the invariant exo description) appears to be blocked. The difficulty has to do with the fact that the observer cannot manipulate his own internal motions in the same easy way in which he can do this with his external state of motion. This puts him or her in an unresolvable quandary — similar to that of the little kitten prevented from actively exploring its world.

One might object that nothing prevents the observer from actively manipulating his own state of internal microscopic motion — for example, by putting his head into the 4-tesla field of a medical magnetic resonance imaging (MRI) machine — with the consequence that a measurable distortion of the outside world should occur (cf. Ref. 7a). The point in the present context, however, is that the observer cannot manipulate his interface in a specific (planned-in-detail) fashion. To do so, he would have to be able in effect to temporarily leave the interface to enter another in order to then compare the two. In the two other known examples of covariance (geometry and hypergeometry), such an alternative vantage point could always be adopted because a "combined interface" also represented an accessible option: two cuts could be combined through communication, either across space (ordinary communication) or across time (self-communication) using memory or a document.

In contrast, an internal-motions-dependent interface cannot be left. For example, it is time-dependent on a time scale so short that it would be impossible to deliberately modify its content in detail even if machinery for micromanipulation (like a Tera waves generator) were available. A more fundamental reason is the absence-in-principle of a record. Records presuppose that the change is not ephemeral. Indeed, no matter what part of the world is looked at (close by or distant, external or internal to the body of the observer, in temporal proximity or deep in the past), it is subject to the difference principle. That is, it will be affected by the momentary state of the observer's internal motions. Since the resulting "distortions" affect the whole world, not even memory provides for an escape. However, any change of the whole world, with nothing left to record the change, is no change. It is imperceptible by definition. Two whole worlds can never be compared since the second would never be complete if residues from the first were recognizable within it.

10.4   Falsifiability

An important counterargument suggests itself at this point: Is Boscovich's covariance at all a falsifiable (and hence scientific) concept? At first sight, a covariance of which never more than a single element is available at a time does not lend itself to detection, since the larger invariance cannot be constructed out of a single covariant building block. The question of falsifiability nevertheless can be answered in the positive. For although the individual distortion remains hidden, its existence is accessible in principle. Otherwise the method of looking at the whole in the form of a model, employed by Boscovich, could never have been embarked upon in the first place.

This method involves picturing an entire universe — so that it is completely transparent to the privileged onlooker. When one is confronted with such a "lower level universe," it is easy to recognize both the existence of the interface and its opacity. Of course, this "toy option" does not prove anything about our own world relative to which we lack a privileged external position. However, the inhabitants of the lower level universe are in principle able to build an even lower level universe of their own and look at it in detail. It would be totally decoupled from their own world. It would be no measuring device. Nevertheless they could by happenstance stumble across a lower level universe which is isomorphic to their own as far as the structure of the interface is concerned. Although they would not immediately recognize this fact, "unreasonable coincidences" would eventually pop into their eyes, linking distortions valid on the lower level with facts known from their own level.

This could cause them to become suspicious and to embark on a systematic program of studying model universes that differ from each other in type and sophistication (like the number of forces incorporated). In this way they would be able to accumulate a catalog of non-exo-objective features of the opaque type that can arise in principle. Although nothing guarantees completeness or even a single overlap with their own world, the inhabitants do have a nonzero chance of correctly identifying some of the hidden laws that govern their own universe — in spite of their being inaccessible. Eventually, only the contiguous individual facts within the identified laws would remain inexplicable. The theory itself, however, would be verifiable, and hence also falsifiable. A program of the same type has been advocated for our own world in Chap. 6.

10.5 Von Neumann's Argument

The fact that Boscovich's covariance is detectable with a nonzero probability if it plays a role in our own world says nothing about its actual existence, in the world. The structure of the cosmos could be such that there is just no room left for Boscovich covariance. In particular, the fact that the principle was discovered in a classical context opens up the possibility that quantum mechanics will thwart its applicability. This question was apparently first raised by von Neumann, in his classic book.[9]

Von Neumann, who was unaware of Boscovich, rediscovered Boscovich's principle in the following form: "The state of information of the observer regarding his own state could have absolute limitations, by the laws of nature." He saw this as a possible way to explain the statistical character of the outcome of a quantum measurement: it could be that "the result of the measurement is indeterminate, because the state of the observer before the measurement is not known exactly." However, von Neumann was able to show in a few lines that in the formalism of quantum mechanis, "the non-causal nature of the process (...) is not produced by any incomplete knowledge of the state of the observer."[9]

This conclusion should not detract from the fact (noted by von Neumann and re-emphasized by Everett[10]) that the unknown state of the observer codetermines the noncausal outcome. Von Neumann only showed that the state of the observer could not have been ascertained independently of the object's state beforehand since the object's state codetermines the state of the observer.[9] It is only in this limited sense that the noncausal nature of the measurement process is "not produced by any incomplete knowledge of the state of the observer." Thus, the Boscovichian explanation remains possible in principle. The interface program is therefore not thwarted by quantum mechanics.

On the contrary, von Neumann's question itself points to a mechanism beyond the quantum world. It is of a classical (Gödelian) type: The "state of information of the observer regarding his own state could have absolute limitations, *by the laws of nature*."[9] This hypothesis states explicitly that quantum mechanics (more specifically, the unpredictable outcome of the reduction of the wave packet) might be explicable by a more fundamental principle (the lack of complete self-knowledge of the observer). This more fundamental principle would have to be explained in terms of a more fundamental theory ("laws of nature") than quantum mechanics itself. Von Neumann's question, when answered only from within the formalism of quantum mechanics, is therefore not answered completely — there remains a "gap" in the proof.

To close the gap will not be easy, since quantum mechanics will have to be transcended in an unknown direction. What is of interest in our context is that this more general question is related to Boscovich's idea. A complete description always implies dependence of the result of any measurement on the state of the observer — according to Boscovich. It is therefore possible to stick to the hypothesis that Boscovich covariance plays a role in our own world — despite the fact that this world is quantum-mechanical.

At the same time the more specific hypothesis can be ventured that it is first and foremost the quantum phenomena themselves which owe their existence to Boscovich covariance. This hypothesis implicitly suggests that the "hyperreality" is classical. Thus, Boscovich's own way of thinking can be re-adopted.

10.6 The Nonlocality Argument

As soon as one speaks of classical hidden variables — as first introduced by Boscovich — one is faced with the objection of nonlocality. Bell[11] showed that one feature of quantum mechanics — its nonlocal nature — can never be explained by a local theory. Boscovich's classical view, however, is local. Nevertheless Bell's counterargument can be shown to be inapplicable with the same ease as von Neumann's.

Bell[11] started out with the presupposition that the nonlocal quantum correlations to be explained are objective in the sense of reflecting a "separable predetermination."[11] Only under this condition are they inexplicable by a local-deterministic hidden variables theory.

This presupposition excluded the possibility that the correlations are interface-generated. In an explanation of quantum mechanics based on Boscovich covariance, the Bell correlations would, while objective for the observer in question, cease to be objective on a directly inaccessible "second level of objectivity." Since Bell[11] was unaware of the distinction between two levels of objectivity (exo and endo), he could not possibly have mentioned it as an exception to his theorem.

Indeed, two objectively equal classical particle motions are distorted in the interface of an explicit observer in the same way. A symmetry assumption of this kind suffices, however, to formally arrive at the Bell correlations.[13] The latter therefore are indeed compatible with a deterministic local theory of classical "hidden variables" of the Boscovich type.

10.7 Observer Relativity

Boscovich's view of the world is reminiscent of Bohr's idea that the observer is a "participator" in the creation of his world (as Wheeler[14] put it). However, another theory in quantum mechanics is even closer in spirit to Boscovich although it is nonclassical — Everett's interpretation.[10] To the internal observer of the universe, there is always only one quantum world ("relative state"[10]) in existence at every moment. In his classic paper on Everett's theory, Bell[12] takes this observer state relativity literally, so to speak: the different Everett worlds, which depend on the momentary (if unknown) state of the observer, exist successively. The ordinary interpretation of their existing simultaneously is thereby discarded. According to Bell's new interpretation of quantum mechanics, the observer jumps back and forth in rapid succession amongst many different, internally consistent Everett worlds.

The shielding principle — which in Everett's formulation takes the place of the postulate of state reduction in the usual formalism — still separates the different quantum worlds, however. Only it does so not laterally (along a new, unphysical dimension) but sequentially (along the time axis). No difference is thereby implied for the observer: at every moment, he finds himself in an intrinsically consistent world with a consistent history. The fact that he (or she) is rapidly taken out of each quantum world and implanted into another is "censored out."

This view of Bell's contradicts common sense. Nevertheless it is less counterintuitive than the canonical interpretation of Everett's theory (in which more than one relative state exists at a time). After uniqueness has been reestablished, all that is lacking in order to make the theory fully acceptable is an explanation for the hopping.

Boscovich covariance provides such an explanation. For it leads to a very similar prediction. The internal motions of the observer "transform" the true modes of existence in an ever-new way — so that the appearance of the world changes as rapidly as the internal motions do. If the latter are microscopic, a similarly rapid change as in the Bell–Everett picture is implicit. Secondly, the changes themselves are imperceptible, too. For if they were perceptible, the interface could be left and compared with another interface — which is impossible, as we saw.

This result is independent of the particular nature of the changes which occur in the interface. This nature is contingent on the particular properties of the classical universe assumed. For example, the question of whether or not Nelson diffusion[15] (and by implication the Schrödinger equation of quantum mechanics) is present depends on those finer details.

A still finer subquestion is whether or not the Nelson diffusion comes with the right constant for the unit action contained in it. Here robustness sets in again: even the second alternative ("nonconstant unit action") reproduces universality. For, at each moment, the distortions of objective reality make sure that the "quantum world" is consistent.

Thus, a Boscovichian world varies imperceptibly along *more* (co)dimensions than the quantum reality described by Bell does. In particular, the value of Planck's constant — and hence the length scale (Bohr radius) — is in general not conserved across worlds. A prediction made by Boscovich is thereby unexpectedly confirmed: not even length scales can be trusted from an exo objective point of view (cf. Appendix).

The two time-*independent* quantum effects already singled out by Boscovich (seeming invariance of length scale; seeming incompressibility) thus acquire time-dependent siblings. They comprise measured eigenstates and everything which can be made dependent on them (like cats, correlations and constants). Boscovich's observer-centered picture in this way appears more and more convincing.

10.8 Main Counterargument

The main counterargument can be lifted from Boscovich's own writings. Toward the end of his paper (Sec. 7 of the Appendix), he has to admit that common sense (the thinking of the "many") is not ready to accept his views, and that even amongst his colleague physicists ("philosophers") he stands completely isolated. There is little doubt that, unless something has changed fundamentally in the attitude of science, the view of Bell presented above (and shown to be compatible with Boscovich's relative state formulation) is as unacceptable today as Boscovich's own views were a quarter-millennium ago.

This preliminary verdict is consistent with the fact that Everett's observer-centered relative state theory is accepted by a minority only. The improvement scored by Bell[11] appears to be unknown. Also, "hidden variables theories" are still out of fashion under the influence of Bell's earlier,[10] much-better-known result. Bohm's[16] hidden variables theory, which is nonlocal, also has found little acceptance despite the fact that the hidden features it introduces are "tame." A full-fledged hidden universe, as introduced by Boscovich, is still harder to accept.

A third point reinforces the counterargument: The above-considered hidden universe is identical in its structure to a theory of the 19th century believed to be outdated in the 20th. This fact too is bound to generate a dizzy feeling.

However, something has changed since Boscovich's time. It is not a new psychology, nor a greater emphasis on experiment, nor an increase in mathematical proficiency, nor the vast increase in the number of scientists — it is the advent of the computer. Only a century ago, it still took a major flight of the imagination to picture an external operator to a kinetic universe. Only a "demon" could tamper with the second law of thermodynamics from the outside. Today, after the invention of the molecular dynamics simulation paradigm by Alder and Wainwright, who were the first to put billiard balls into a computer,[17] any connotation of sorcery has long vanished.

10.9   Second Level Experiments

It is not the computer alone which makes the new type of explanation acceptable today. Nor is it the fact that with the aid of this instrument, new predictions regarding the world can be arrived at. The decisive new element rather is that we are ready to test these predictions.

The notion of experiment was "invented" during the Renaissance. Although the Greeks already used it with great mastery, its most important function — to lead the way out of the darkness of speculation — remained undiscovered in antiquity. To blame is the subconsciously held prejudice that experimentation is "childish" (or, synonymously, "slavelike"). A closely related prejudice has kept theory in its grip until recently: Theory had to be "subservient" to experiment. That is, any theory-based motivation for an experiment had to be "serious." Again there was no room left for playfulness. Second level experiments would have been frowned upon, not only by the ancient Greeks, but also by the contemporaries of Boscovich's and Maxwell's and even in the age of mainframes still.

A second level experiment (or meta-experiment) is an experiment that, although performed in the real world, takes its departure from the level-jumping paradigm (the exo/endo distinction introduced by Boscovich). The underlying rationale is the discovery (Sec. 10.4) that the inhabitants of a toy universe have access in principle to acts whose outcome would teach them facts about their own world that are ordinarily hidden to them. Of course, to embark on these very acts or actions, the inhabitants would first need to be "cued." The only thing that can cue them is a still lower level universe.

From the point of view of traditional (serious) theory, it would be irresponsible to take this analogy at face value in our own world and start experimenting right away. Probability stands in the way. More specifically, the nonuniqueness principle (Sec. 10.4) lets the approach appear hopeless from the beginning. Too many model universes seemingly need to be investigated as to the properties possessed by their interfaces. The endeavor therefore has the character of "mathematical playwork."

On the other hand, nothing stands in the way of one's starting to experiment right away. Only an "adult" would not even try because of being too well aware of the pitfalls. The situation thus is the same, in a sense, as it was at the time of the ancient Greeks. The point is that it costs nothing to try. This easy attitude toward experiment is only a product of the age of the personal computer, which lets "having the cake and eating it" no longer appear impossible.[18] Explanation, prediction and venture, taken together, form a new triad at the dawn of the information age. Second level experiments are there to be tried.

10.10   An Example

Let us not be too daring at this point. Since it is too early to get a concrete hint from a particular lower level universe, all that can be done with confidence today is exploit the fact that the outlook on our own world has changed. New questions concerning standard experiments come into view that would have been impossible to consider outside the toy universe paradigm.

There exists an experiment which combines Einstein relativity and Boscovich relativity. Ideally, one would like to first have a relativistic molecular dynamics simulation to back everything up with serious computer work, but then there is no relativistic molecular dynamics possible because of the "no interaction theorem."[18a] (A gas with long-range forces acting between the particles cannot be described from the point of view of a moving frame.[18a]) On the other hand, this does not matter in the present case because there is virtually no disagreement about the outcome of the experiment anyhow. In fact, the agreement is so unanimous that the experiment will perhaps never be done even though it is feasible with current technology.[19]

The problem was apparently first glimpsed by Susan Feingold in unpublished 1978 notes (quoted by Peres[20]). It is a standard "correlated photons' experiment in the sense of Bell — with a minor change. The two measuring stations (consisting of analyzer, detector *and* observer) do not both belong to the same inertial frame but, rather, recede from each other. The idea is to have each side reduce the singlet-like superposition state of the photon pair first in its own frame. A very simple space–time diagram — a light cone crossed by two slanted planes ("VX diagram," where "V" stands for the light cone) — shows that this is possible. We suppose that current theory (cf. Ref. 20) is correct in its predicting a negative outcome ("no change of correlations" compared to an ordinary Bell type experiment) and ask the question: Why does this commonly expected outcome amount to a proof of observer-centeredness?

The expected outcome implies that each of the two observers can stick to the assumption that the other's measurement was done only after the singlet state had already been reduced by himself or herself. For the photon impinging on the other observer's measuring device at that time already possessed a fixed spin in the frame of the first. Hence the subsequent registration of that spin only revealed exactly the properties to be statistically expected from a photon having this particular pre-existing spin (since, according to relativistic

quantum mechanics, a receding measuring device always measures an unchanged photon spin, so that the logic of a common stationary frame is indeed
applicable). Thus there is no difference compared to an ordinary Bell experiment — for each of the two observers. For there is complete agreement with
what each expects.

The two interpretations held "symmetrically" by the two observers are at
first sight "compatible." The Bell correlations are, as is well known, the same
no matter which side reduces the singlet state by making the first measurement.[10,20] In an ordinary (single frame) Bell experiment, each side can therefore pretend to bear the responsibility for the other's correlated outcome.[20a]
However, since the two observers are mutually at rest, it is nonetheless always
possible in principle to find out which side accomplished the reduction. (Passing over to a different reference frame could change this interpretation, but
not without additional hypotheses.) Therefore, no more than one side can be
exempted from being responsible for the reduction. In the present case, in
contrast, this exemption can be demonstrated to each side, by the other. This
clinches the case.

Still, since each of the two allegations, taken alone, is compatible with the
observed facts, it looks as if the two taken together were also compatible. This
is not the case, however. Although at first sight everything looks all right since
the world of each observer is subject to the rules of nonrelativistic quantum
mechanics in accord with Bell's theory, something is subtly wrong. One cannot
get rid of the impression that there is "too much" symmetry involved. This is
exactly the case.

A similar situation has been encountered once before in the history of
physics. At that earlier time, no one took offense until a whole new branch
of physics had to be created as a remedy. After the motion-independent con-
stancy of the speed of light had been discovered experimentally by Michelson
and Morley, and Lorentz and Fitzgerald had formulated the contraction hypothesis which "explained" the phenomenon, ten more years had to pass before
the decisive question was put forward by Einstein in 1905: Should not each of
the two observers be alarmed that it is always the other side which contracts
so conveniently?

Einstein realized that too much symmetry is a valuable cue. Whenever two
observers, when taken alone, have a consistent description of the world each,
but not when taken together since the two descriptions contradict each other
(for the same facts are explained differently by the two sides even though both
make use of the same theory), then this is an example of "too much symmetry."

Such a paradox is an empirical hint that a new covariance principle waits to be introduced.

As soon as one has been alerted to such an empirical state of affairs in a concrete case, it is necessary to search for the deeper — invariant — picture from which the two parallel but incompatible (covariant) phenomena must have sprung. In the present case, there are two ways to arrive at an invariant solution.

The first possibility is to postulate that quantum events are "cooperative" events: the future — the other observer — participates in the creation of the presently observed outcome on an equal rights basis. This is the message taught by field theory and relativistic quantum mechanics. A nonuniqueness (overdetermination[19]) in space-time allows the observer in each frame to pick the facts that in this frame are consistent with nonrelativistic quantum mechanics. The commutator relations can then be violated, although so only in retrospect (which according to Heisenberg is admissible); cf. Ref. 20a.

The alternative second possibility departs from Bohr's view that the decision on the part of the observer as to what to measure critically determines how the world is going to look. All other observers, linked to him by linguistic communication, are bound to reach the same result afterwards. Everett's interpretation[10] says exactly the same thing — confirmation of the result obtained by the first in his own world by the others. Nevertheless, there is an added element of "free play" present which ordinarily is considered redundant, since predictive isomorphism to Bohr's world suffices under all conditions known so far. This added element acquires crucial significance in the present case. Although each Everett world is internally consistent (and indistinguishable from a Bohr world), one would be mistaken to assume that the Everett worlds of two observers are equal. Not even two Everett worlds of the same observer are equal.

This is the reason why, unlike Bohr's quantum mechanics, Everett's can be brought into a Boscovich-covariant form. In analogy with Einstein's solution in the earlier case ("each side complies" — is contracted — in the frame of the other), it is again possible to say "each side complies" — is perturbed — in the world of the other. Just as in the previous case, the existence of more than one "frame" (with the same laws but deviating measurement results) had to be acknowledged, so in the present case the existence of more than one "world" (with the same laws but deviating measurement results) has to be acknowledged.

The existence of more than one Everett world can thereby be demonstrated. The mentioned consensus that the experiment need not even be done (because the outcome is taken for granted) is tantamount to accepting one of two facts: either the commutator relations of quantum mechanics can be violated, or the existence of more than one quantum world is accepted. The latter conclusion is in accord with a prediction made by Boscovich's second relativity principle.

If the first of the two alternative conclusions is adopted, the invariant reality can be indicated immediately (although it turns out to be overdetermined). It consists of the usual space–time of quantum field theory which is nonunique since quantum eigen states are frame-dependent.[20b] If the second solution is adopted, the invariant absolute world has yet to be found. None of the currently discussed hidden variables theories (including stochastic ones[21]) appears to be up to the "enlarged" task of accommodating more than one quantum world. The only candidate theory is the above proposal. Thus an empirical reason why a theory as unfamiliar in structure as Boscovich's may be needed appears to have been found.

10.11 Conclusions

A new "rationalistic" theory of the physical world has been presented. It revives a proposal made almost 250 years ago by Boscovich.[28,29] The idea stands in a long tradition in science which includes the names of Heraclitus and Maimonides, to mention only two. The true reality is different from the way it appears. Nevertheless it is possible to find out about this fact, such that some — or even all — of the laws involved can be discovered.

The approach in a sense generalizes Gödel's[22] "valid from the inside" limitation principle. Unlike previous attempts to capitalize on Gödel's mathematical insight for physics, it looks at explicit observers in a continuous reversible context. It still suffers from the drawback that fields, even though eventually needed to explain the discrete spectrum of particle types, are not included. Therefore a complementary program like Wheeler and Feynman's[23] which allows one to get rid of the infintely many degrees of freedom of a classical field, will be needed in the long run.

Boscovich recognized properties of our own (quantum-governed) world as incompatible with the consistent classical picture developed by himself. To account for these static properties (fixed length; incompressibility), he was compelled to search for a new type of explanation. He found it in a dynamic argument — that of the "breathing" world (see Sec. 2 of the Appendix). If one takes the theory of the "time-dependent interface" (implicit in Boscovich's approach) seriously, it is apparently possible to arrive also at a qualitative explanation of the three major time-dependent phenomena of quantum mechanics — stochasticity, nonlocality and state reduction.

What is open is whether quantitative details (like the numerical value of Planck's constant and its degree of isotropy in space) can also be captured eventually. A precondition for tackling such questions is complete success in the task of calculating and visualizing the simplest case. How is a third particle which acts as an environmental forcing "seen" by a chaotic subsystem made up of two (and then three and four, and so on) equal particles if all necessary symmetries are taken into account? Afterwards, T, the mean interval between effective time reversals in a multiple-particle explicit observer, will become calculable, and from it, the interface-specific elementary action, ET

However, even more important than qualitative explanation and quantitative prediction of facts already known may be the fact that the present approach alerts us to new possibilities. New covariances may exist empirically in which the observer plays a privileged role.

In this context let us consider the most extreme observer-centered phenomenon possible: the phenomenon of nowness. Bell was the first to realize[12] that the sealing-off principle employed by Everett[10] has the same structure as the sealing-off that occurs phenomenologically in the transition from one momentarily valid world to another. Subjective nowness was thereby elevated from the status of a codimension 1 phenomenon to that of a codimension 2 phenomenon (with no longer only the time axis but also Everett's world-determining parameter participating in the unfolding). However, no rationale for this elegant hypothesis could be offered.

The present approach based on Boscovich covariance not only unexpectedly confirms and explains Bell's version of Everett's theory, as was shown above, but at the same time explains "nowness." Since the previous states of the interface are not represented within the interface itself, they are extinguished, in effect, along with everything they entail, from one time slice to the next. Hence there exists a "cross-inhibition" between neighboring elements on the time axis that has no corollary on the other (spatial) axes. The present approach thus explains more than it was meant to. The formulation of nowness as a covariance principle unexpectedly arises.

Nowness on the other hand belongs to subjective phenomenology. So far, only macroscopic brain states have been known to possess subjective correlates. For example, the microscopic states are irrelevant for the macroscopic workings of a computer. Could they nevertheless be relevant to the "world" of such a machine, should the latter possess a world? The interface principle suggests that the answer is yes.

Does this mean that subjective experience may perhaps become predictable scientifically? The answer is no. We can never hope to find anything more than objective correlates. On the other hand, a theory admitting an observer-private interface into the picture may allow much finer correlates to be found than has been possible before.[7a]

The present approach therefore in a sense revives Cartesian rationalism. According to Descartes, rationalism would be dead (falsified) if a single instance of sorcery were reliably reported.[25] Equivalently, it would be sufficient if a single example of an in-principle-inexplicable arbitrariness were found to reside within the world. This catastrophe is generally believed to have occurred with the advent of quantum mechanics. A case in point is a Dehmelt atom.[26] Such an immobilized single atom emits (if subjected to two laser beams of the right frequencies and intensities) a Morse-code-like time signal of light pulses visible to the naked eye. It is one of the simplest quantum systems, although its theory up to numerical simulation is rather involved.[26a] Methods of chaos-theoretic time series analysis can be applied to it (whereby novel dimension problems are expected to arise[26b]). This "telegraph noise of nature" is, nevertheless, in its actual structure not described by quantum mechanics. Only the probabilities of the individual state transition events belong to the subject matter of modern physics. Hence the investigation of any concrete Dehmelt signal is useless according to modern science: to have a look through this "telescope" is discouraged.

Any particular sequence of symbols generated by the atom has the same interest as a paranormal phenomenon has. The inexplicable by definition not only invites sorcery, it is sorcery. Would it make a difference if a sorcerer of high standing — like Don José,[26c] Castaneda's teacher (better known as Don Juan) — were able to prove that he could influence the code? To exempt a part of relational phenomenology from causal explanation — as quantum mechanics has done — is tantamount to giving up rationality in favor of superstition.

Boscovich's approach combines Einstein's proposal to look for a mechanistic explanation with Bohr's suggestion to look for a holistic one. The "bad dream" that both insisted would pertain in the absence of a rational explanation would thereby be avoided. However, would the price to pay — a return to old-fashioned Cartesian rationalism — not be too steep? The answer appears to be: On the contrary.

Cartesian rationalism consists of a single falsifiable hypothesis: The world, which to the individual has the character of a Big Dream, may be well defined as far as the quantitative relations between its qualitative features are concerned (hypothesis of relational consistency). The universe then would behave like a giant dynamical system ("machine"). The first major implication is determinism, while the second implication is exteriority.

Determinism is sometimes considered undesirable because it seems to imply the absence of free will. This is incorrect according to Descartes. For one particular partial machine would be exempt. Although the dynamical states of this machine would lawfully reflect the content of the whole Dream, they could not be said to determine it. For the machine in question — the brain of the observer — would be an "internal character" of the dream. Nevertheless it remains true that the brains of all other individuals would indeed be "just machines" in the first's Dream. To endow them with free will, too, requires a further step. One possibility would be an Everett-like new dimension giving the other players more freedom than they have in the world of the first. (Compare this with the discussion of an added dimension by Boscovich,[5] p. 234, bilingual edition of 1922.) A simpler possibility is the deistic assumption of "preadapted" initial conditions (just as this was already assumed for the dreamer's brain). The real problem which poses itself, therefore, is not metaphysics but ethics.

The exteriority principle was invented by Descartes in the 17th century. It was elaborated by Ludwig Feuerbach in the 19th and Martin Buber and Helmut Plessner in the 20th. It was rediscovered and named by Levinas.[27] Lower level worlds (in the computer) by definition pose an ethical problem because of one's being completely exterior to them. However, this "vertical exteriority" is not fundamentally different from the "horizontal" one (between individuals) implicit in a consistent world. If the other individuals are only machines within the first's Dream, the dreamer is equally exterior towards them — and hence privileged — as toward a lower level universe. Absence of sorcery paradoxically conveys the ultimate power. This was a desirable state of affairs to Descartes. He felt that it made the arbitrariness of the whole dream acceptable.

If the victim of the dream cannot rule out, within the dream (due to the apparent machine-like character of the dream's internal relations), that he is endowed with as much power towards others as is brought to bear against him, he is even. For he (or she) can then refrain from misusing the infinite power.[27a] The unlimited power of exteriority can be used to create one thing even if it has never existed before: fairness. Rationalism thus is unexpectedly a most desirable option. It had to be created anew by philosophy after science had abandoned it.

The present attempt at restoring rationalism in science is therefore admissible. Many details are still wanting before the attempt can be called a success. The alternative outcome will be that quantum mechanics gets confirmed as the most fundamental theory of nature possible despite its irrationality. The greater the vigor of the attempt to falsify the rational Cartesian–Boscovichian position, the greater the chances that a reliable position will be found.

To conclude: a scientific idea proposed in 1755 by Boscovich and rediscovered in 1932 by von Neumann has been re-examined. No fault could be found with it. Boscovich's proposal belongs to a class of physical principles whose independent stance has become fully visible only today: covariance principles. Boscovich covariance is the most general example discovered so far. It is so general, in fact, that it ceases to be demonstrable directly. Maybe it is only a dream. Alternatively there exists a new key to decode the world.

Acknowledgments

I thank John Casti, Anders Karlqvist, Karl-Erik Eriksson, Peter Landsberg, Bob Rosen, Itamar Procaccia and Brian Goodwin for stimulation. Discussions with John Bell, John Wheeler, Joe Ford, Giulio Casati, Jim Yorke, Ray Kapral, Mike Mackey, Leon Glass, Hans Troger, Leon Chua, Pauline Hogeweg, Peter Gray, Arnold Mandell, Peter Arhem, Stig Andersson, Richard Goodwin, Günter Mahler, Boris Schapiro, Günter Hoff, Peter Erdi, George Kampis, Klaus Kahlert, Martin Hoffmann, Peter Weibel and Jürgen Parisi were helpful. Werner Kutzelnigg first mentioned Ruder Boscovich's name to me in 1988. Jens Meier dug up the 1758 textbook and contributed to the discussion of the double-frame experiment. The 1989 Abisko conference under the midnight sun acted like a focus.

References

1. R. Held and A. Hein, "Movement-produced stimulation in the development of visually guided behavior," *J. Comp. Physiol. Psychol.* **56**, 872–876 (1963).
2. E. F. Taylor and J. A. Wheeler, *Spacetime Physics* (Freeman, San Francisco, 1963), p. 42.
3. H. Poincaré, *La Valeur de la Science* ("The Value of Science") (Flammarion, Paris, 1904), p. 98.
4. A. Einstein, *Zur Elektrodynamik bewegter Körper* ("On the electrodynamics of moving bodies"), *Ann. der. Phys.* **17**, 891–921 (1905).

5. R. J. Boscovich, *De spatio et tempore, ut a nobis cognoscuntur* ("On space and time as they are recognized by us"), 1755. Reprinted in his *Theoria Philosophiae Naturalis* ("Theory of Natural Philosophy"), Vienna, 1758, Supplement II. See the Latin English edition of the second Venetian edition, 1763: *A Theory of Natural Philosophy* (J. M. Child, ed.) (Open Court, Chicago, 1922), pp. 404–409. The English part was reprinted by MIT Press, Cambridge, 1966; see pp. 203–205.

6. H. G. Alexander, *The Leibniz Clarke Correspondence* (Manchester University Press, Manchester, 1956), pp. XLIV–XLV.

7. S. G. Brush, *The Kind of Motion We Call Heat — A History of the Kinetic Theory of Gases in the 19th Century*, Book 2, pp. 594–595 (North-Holland, Amsterdam, 1976); The passage with the demon, copied by Laplace without mentioning the source, is: Ref. 5, pp. 141–142, Latin–English edition. Boscovich's authorship was discovered by K. Stiegler, *Proceedings of the 13th International Conference on the History of Science* (Moscow, 1974), p. 307.

7a. O. E. Rössler and R. Rossler, "Is the mind–body interface microscopic?", *Theoretical Medicine* **14**, 153–165 (1993).

8. O. E. Rössler, "Endophysics," in *Real Brains, Artificial Minds* (J. L. Casti and A. Karlqvist, eds.), (Elsevier, New York, 1987), pp. 25–46; cf. Chap. 6 of the present book.

9. J. Von Neumann, *Mathematical Foundations of Quantum Mechanics* (Princeton University Press, Princeton, 1955), p. 438; first German edn., 1932, p. 233.

10. H. Everett III, "'Relative-state formulation' of quantum mechanics," *Rev. Mod. Phys.* **29**, 453–463 (1957).

11. J. S. Bell, "On the Einstein–Podolsky Rosen paradox," *Physics* **1**, 195–200 (1964).

12. J. S. Bell, "Quantum mechanics for cosmologists," in *Quantum Gravity 2* (C. J. Isham, R. Penrose and D. Sciama, eds.) (Clarendon, Oxford, 1981), pp. 611–637.

13. M. Hoffmann, *Fernkorrelationen in der Quantentheorie: Eine neue Interpretation* ("Correlations-at-a-distance in quantum theory: a new interpretation"), PhD thesis, University of Tübingen, 1988; "A local-realistic explanation of EPR correlations," *Found. Phys.* **20**, 991–996 (1990).

14. J. A. Wheeler, "Beyond the black hole," in *Some Strangeness in the Proportion — A Centennial Symposium to Celebrate the Achievements of Albert Einstein* (H. Woolf, ed.) (Addison-Wesley, Reading, 1980), pp. 341–375.

15. E. Nelson, "Derivation of the Schrödinger equation from Newtonian mechanics," *Phys. Rev.* **150**, 1079–1085 (1966).

16. D. Bohm, "A suggested interpretation of the quantum theory in terms of 'hidden variables,'" *Phys. Rev.* **85**, 166–179, 180–193 (1952).

17. B. J. Alder and T. E. Wainwright, "Studies in molecular dynamics," *J. Chem. Phys.* **27**, 1208–1209 (1957); *Scientific American*, Oct. 1959.

18. T. A. Bass, *The Eudaemonic Pie* (Houghton-Mifflin, Boston, 1985); 2nd edn.: *The Newtonian Casino* (Longman Group, London, 1990).

18a. E. H. Kerner (ed.), *The Theory of Action-at-a-Distance in Relativistic Particle Dynamics* (Gordon and Breach, New York, 1972); the term "no interaction theorem" is due to Herbert Goldstein: H. Goldstein, *Classical Mechanics*, 2nd edn., pp. 332, 334, 362 (Reading, Addison-Wesley, 1980).

19. O. E. Rössler, "Einstein completion of quantum mechanics made falsifiable," in *Complexity, Entropy and the Physics of Information* (W. H. Zurek, ed.) (Addison-Wesley, Redwood City, 1990), pp. 367–373.

20. A. Peres, "What is a state vector?," *Am. J. Phys.* **52**, 644–649 (1984); cf. also: S. Feingold and A. Peres, *J. Phys.* **A13**, 3187 (1980).

20a. O. Costa de Beauregard, "Lorentz and CPT invariances and the Einstein-Podolsky-Rosen correlations," *Phys. Rev. Lett.* **50**, 867–869 (1983).

20b. Y. Aharonov and D. Z. Albert, "Is the usual notion of time evolution adequate for quantum-mechanical systems? II. Relativistic considerations," *Phys. Rev.* **D29**, 2228–234 (1984).

21. F. Guerra and M. E Loffredo, "Stochastic equations for the Maxwell field," *Lett. Nuovo Cim.* **27**, 41–45 (1980).

22. K. Gödel, *On Formally Undecidable Theorems* (Basic Books, New York, 1962), first published in German in 1931. See also: P. Weibel and E. Köhler, "Gödel's undecidability proof: contours in the history of ideas of a famous mathematical theorem" (in German), in *Gödelsatz, Möbius-Schleife, Computer-Ich* (F. Kreuzer, ed.) (Franz Deuticke Verlagsgesellschaft, Vienna, 1986), pp. 72–101.

23. J. A. Wheeler and R. P. Feynman, "Interaction with the absorber as the mechanism of radiation," *Rev. Mod. Phys.* **17**, 157–162 (1945).

24. O. E. Rössler, "Explicit observers," in *Optimal Structures in Heterogeneous Reaction Systems* (P. J. Plath, ed.), pp. 123–138 (Springer-Verlag, Berlin, 1989); cf. Chap. 8 of this book.

25. R. Descartes, *Meditationes de Prima Philosophia* ("Meditations on the First Philosophy") (Soly, Paris, 1641).

26. W. Nagourney, J. Sandberg and H. Dehmelt, "Shelved-optical-electron amplifier: observation of quantum jumps," *Phys. Rev. Lett.* **56**, 2797–2799 (1986).

26a. W. G. Teich and G. Mahler, *Phys. Rev.* **A45**, 3300 (1992).

26b. T. Erber and S. Putterman, "Randomness and quantum mechanics — nature's ultimate cryptogram?," *Nature* **318**, 41–43 (1984); *Archives Phys.* 190, 259–309 (1989).

26c. P. Das (P. C. Adams), "Initiation by a Huichol Shaman," in *Art of the Huichol Indians* (K. Berrin, ed.) (Harry N. Abrams Publishers, New York, 1978), pp. 129–141; p. 212.

27. E. Levinas, *Time and the Other* (R. A. Cohen, transl.) (Duquesne University Press, Pittsburgh, 1987); French original 1946, first French edn. 1947. Cf. also: L. Wenzler, *Zeit als Nähe des Abwesenden* ("Time as closeness of the absent"), postface to the German edition, *Die Zeit und der Andere* (Felix Meiner-Verlag, Hamburg, 1984), pp. 67–96.

27a. E. Levinas, *Totalité et Infini, Essay sur l'Extériorité* ("Totality and Infinity — An Essay on Exteriority") (Martinus Nijhoff, The Hague, 1961).

28. E. Heilbronner, "Why do some molecules have a symmetry different from that expected?", *J. Chem. Educ.* **66**, 471–478 (1989). See also: W. Johnson, "The shadowy figure of R. J. Boscovich (1711–1787)," *Int. J. Mech. Sci.* **33**, 579–591 (1991).

29. L. L. Whyte (ed.), *Roger Joseph Boscovich — Studies of His Life and Works on the 250th Anniversary of His Birthday* (Allen and Unwin, London, 1961); Z. Dadić, *Ruder Bošković* (Yugoslavian-English text) (Skolska Kujiga, Zagreb, Masanykova 28, 1987); for a contribution on Boscovich by Niels Bohr, cf. *Actas des Symposiums R. J. Bošković 1958* (Comitá Inter-académique R. J. Bošković, Belgrade–Zagreb–Ljubliana, 1959).

Appendix

On Space and Time as They Are Recognized by Us
by R. J. Boscovich (1755)[5]

1. We cannot recognize absolutely the local existence modes nor distances nor magnitudes.

In the preceding paper ("On Space and Time"), we have dealt with space and time as they are in themselves. It remains for us to discuss — if we are to say something of lasting value — how space and time are recognized. We by no means immediately recognize through our senses the true (real) existence modes nor can we distinguish one from another.

It is true that we do perceive, from the discriminate impressions (ideas) excited in our minds by the senses, a distinct relation of distance and [reciprocal] position, generated by a pair of local existence modes. However, the same impression can be brought about by innumerable pairs of such modes which — as true points of location — induce relations of equal distance or similar position, both relative to each other and relative to our own organs and the other surrounding bodies.

Two points of matter, for which a distance and [reciprocal] position were induced by two specific existence modes at a first place, may acquire a relation of equal distance and similar mutual position under the influence of two other existence modes at a second place — whereby the two distances will be parallel to each other. If now these two points, as well as we ourselves and all the surrounding bodies, changed their true positions in such a way that all distances remained equal and parallel, we would still have the same impressions.

And we would also have the same impressions if, under preservation of the magnitudes of the distances, all directions were rotated by an equal angle while all mutual angles are the same as before.

And even if the distances were shrunk, under preservation of the angles and mutual proportions and with also the forces staying unaffected by the change of distance while their magnitudes — that function (curve) whose ordinate yields the force — were changed proportionally, we would still undergo no change in our impressions.

2. A motion which is common to us and the world cannot be recognized by us — not even if the world as a whole were increased or decreased in size by an arbitrary factor.

From the above it follows that if this whole world which lies before our eyes were moved forward in a parallel motion in an arbitrary direction (plane) and at the same time rotated by an arbitrary angle, we could not sense this motion and rotation. Similarly, if this whole line of flight — the room in which we are and the plains and the mountains outside — were by an arbitrary motion of the earth rotated in the same sense, we could not sense such a kind of motion. For impressions equal to the senses would be excited in the mind.

It even is conceivable that this whole world before our eyes contracted or expanded in a matter of days — with the magnitude of the forces contracting or expanding in unison. Even if this occurred, there would be no change of the impressions in our minds and hence no perception of this kind of change.

3. Whether it is our own position which changes or that of all the things we see makes no difference for our impressions — so that we can legitimately attribute the motion, neither to ourselves nor to the rest.

If either the external objects or our own organs change their true existence modes in such a way that the above equality or similarity conditions are no longer fulfilled, then indeed the impressions are altered and we do perceive the change.

However, the [changed] impressions are exactly the same, no matter whether it is the external objects that suffer the change or our own organs — or both to an unequal degree. Always our impressions represent the difference between the new state and the previous one, never the absolute change that occurred since it does not fall under the senses. No matter whether the stars are moved about the earth or whether the earth with us on it turns around in the opposite direction, the impressions are the same, the sensations are the same.

The absolute changes we can never sense, but the difference (discrimination) from the previous form we do perceive. If nothing admonishes us of the change of our organs, we always believe ourselves to be unmoved — in accord with the common saying that we take for nothing that which is not in our minds since it is not recognized. Thus, *we attribute the whole change to objects that are situated outside ourselves.* In this way a person errs who, enclosed in a ship, judges himself to be motionless but the shore and the mountains and even the water wave to be in motion.

4. Although we judge the equality of two things from their being equal to a third, we never have a congruence of length in space or time that would not be indirectly inferred.

It is important to note that from the above principle of invariance (unchangeability) of those things whose change we do not recognize through our senses, also the method is derived which we use when comparing the magnitudes of intervals. Thereby we take what we use as the meter to be invariant. We adhere to the principle that "two which are equal to the same are equal," with its two corollaries, "two which are equal multiples or submultiples of a third are also equal to each other," and "two which are congruent are equal."

Let us take the specific example of a ten-foot ruler made of wood or iron. If we find it congruent to some interval — after having applied it once or a hundred times, respectively — and then congruent to another interval — after having applied it once or a hundred times, respectively — we call the two intervals equal.

Furthermore we take that wooden or iron ten-foot ruler to be identically the same standard of comparison after the translation in space. If the ruler consisted of perfectly continuous and solid matter, it could indeed be taken to be the same standard of comparison. However, it follows from my above theory concerning the mutual distance of points that in truth, all points of that ruler continually change their distances during the translation. For the distance is constituted by the true existence modes — which change all the time.

If the change were such that the successive modes, before and after the translation, established true relations of equal distance, then the standard of comparison would, while not identical, at least be equal in length — and the equality of the measured intervals would be correctly inferred. However, we have no way to directly bring together for the sake of comparison the length of the ruler shown at the first place, where it is constituted by the first pair of true modes, with the length of the same ruler shown at the second place, where it is constituted by the second pair of true modes — just as we are unable to bring those intervals together directly.

Since we perceive no change during the translation, as far as the relation of length is concerned, we take the length to be the same. In truth, however, the length is always not unsubstantially changed during the translation. It could even happen that the ruler underwent a giant change, both the ruler and our senses, without our perceiving it — only to resume, after the return to the former place, a state equal or similar to the former.

A small change will nevertheless occur in any case. For the forces which connect the material points among each other must, under a change of position relative to all the other points of the world, suffer a non-negligible change. The same thing by the way also holds true within the usual theory. For no material body is free from little free spaces inside such as to be totally incapable of any compression or dilation. Therefore it is generally believed that at least a small dilation or compression accompanies every translation. Nonetheless we take the meter to remain the same since — as I have pointed out — we feel no change.

5. It is concluded that common sense and the judgment of scientists are at variance.

The consequence of all this is that we can neither recognize absolute distances directly in any way, nor compare them with each other by means of a common standard. We have to estimate their magnitudes through the impressions by which we recognize them, and we have to use meters as common standards that according to common sense (popular opinion) have undergone no change.

The physicists (philosophers), however, have to acknowledge the occurrence of change. Since they see no reason for a marked inhomogeneity they assume the change to occur homogeneously (equally).

6. Even though the true modes which bring about the relation of distance (interval) do change during the translation of the ten-foot ruler, the intervals themselves may under certain conditions be taken to be equal.

Although in truth, distance changes whenever points of matter change their place like the ten-foot ruler, since the true modes which constitute those points have changed, it is nevertheless possible to adopt the following convention of speaking:

- If the change takes place in such a way that the later distance is equal to the previous one, we call the distance "the same distance" and "unchanged."
- If two distances on the same meter (standard) are equal, we call them "the same distance."
- If a magnitude is defined by such equal distances, we call it "the same magnitude."
- If two directions are parallel, we call them "the same direction."
- If neither the magnitude of two distances nor their parallelism has changed, we call them an "unchanged distance-cum-direction."

7. The same reasoning applies to time. Here, even common sense knows that the same interval cannot be transposed for the sake of comparison — although it erroneously denies the same facts about space.

What we have said about the measurement of space is not hard to transpose to time. In the case of time too, we lack a safe and constant meter (measure). No matter how much we may try, we will never succeed in obtaining a perfectly uniform motion. More about this topic — and about the nature of our impressions and their succession in time — is to be found in my [lecture] notes. I only add one point here. When it comes to measuring time, not even common sense thinks it possible that the same interval (measure) of time may be translated from one temporal position to another. Common sense realizes that the interval (measure) is a different one but it assumes that it is equal because of a presupposed uniformity of translation.

In spatial measurement it is according to my theory no less impossible than in temporal measurement to transpose a certain length interval — or time interval (duration), respectively — away from its seat to a new one, in order to have a comparison of the two by means of a third. In both cases another length — or another time interval, respectively — is substituted which is thought to be equal to the first. However, new true positions of the points of the same ruler constitute the new distance, and a new circle is drawn with the same pair of compasses, and a new temporal distance lies between the second pair of initial and final point, respectively.

In my [book] *Theory*, the same perfect analogy between space and time are employed. Common sense believes that in spatial measurement, [identically] the same standard of comparison can be used. Also almost all other physicists (philosophers) believe that, at least in the case of a perfectly solid and continuous meter, one can use the "same" standard in space, but in time only an "equal" standard. I, however, in both cases opt for "equality" only, never for "sameness."

[*Translator's note:* The original numbering of the 7 sections of the above paper was "18–24" in continuation of the 17 sections of the paper preceding it entitled "On Space and Time." The two 1755 papers thereby form a twin set: Physics from without — paper I — and physics from within — paper II.]

11
Can There Be Two Chains
of Causality?

Abstract

The distinction between "tachyons" (faster-than-light particles) and "bradyons" (slower-than-light particles) is re-examined from the model universe point of view. While the former — superluminal — set of solutions in relativity has always been considered pathological, since it seems to require an imaginary mass, the possibility of a more down-to-earth class of tachyons existing has never been ruled out, it appears.

A new suggestion is implicit in the "endo" approach to physics. According to the latter theory, the true (exo) reality is accessible only through the "interface" between the observer and the rest of the universe. Therefore, nothing guarantees that an object in the interface will be represented, (a) "undistorted" and (b) "unique." The first feature (distortion) is familiar in endophysics; the second (nonuniqueness) appears to be new. It enables the following more specific question to be asked: Is the nonuniqueness — if it exists — recognizable from the inside of the world?

A model universe can be used to obtain a first answer. It is the standard classical endophysical universe — with the added feature that the internal observer has a fixed diameter (S). This diameter refers, not to the whole "body" (including supporting and measuring machinery) but to the "observing subsystem" inside. The latter may consist either of a single "sphere" of mean diameter S, or of many such spheres ("grape-shaped" or multicellular observer). In addition, it is assumed that all excitations elicited within S (the same grape) are equivalent for the observer. Apart from these two added assumptions, the usual endophysical findings (micro-time-reversals of mean period T, in the observer, perturbation energy E, in the observer) are supposed to hold good.

S and T, combined, formally yield a velocity (S/T). The latter can be interpreted as a limit to the observable speed of by-flying objects. This follows because any causal influence which, via some measuring chain, reaches the observer from some point x on a by-flying object, necessarily returns to within S during the next time slice, T. This "return" is independent of how fast the object moves, exo-objectively speaking. Therefore seemingly no point of an external object can "pass by" faster. Moreover, since a third point of the object is mapped into S during T, objects that are exo-objectively faster than S/T will seemingly be "shrunk" in their forward direction. In this way, a "qualitative analogy" to relativity (which also combines a speed limit with a shrinking rule) applies.

The "curbed" velocities would correspond to the standard "subluminal" motions (bradyons) of relativity. However, these motions would represent only one type of macroscopic causality accessible to the internal observer of the model universe. A "second solution" (of a "superluminal" type, so to speak) cannot be ruled out to make itself felt also in the world of the internal observer of the artificial universe. While the whole causal structure would be governed by the first ("endo") solution as we saw, individual exo events violating this fabric can still be registered by the internal observer. The question is whether the observer can be given an appropriate "astrolabium" (computing device) which would enable him or her to find out about the underlying correlations. That is, a "second causality" may exist in the model universe and the internal observer may be able to embark on experiments designed to expose the nonuniqueness of causality.

The question which poses itself to us is whether or not the same option — to test for a second causality of superluminal type — is open in our own universe. This presupposes that some of the assumptions made about the artificial universe are related to the exo properties of our own reality, which are unknown.

Acknowledgments

I thank Alfred Huebler, Vilem Flusser, Anthony Freeman, Raima Larter and Hava Siegelmann for discussions. For J. O. R.

12

The Golden Thread of Paradise — An Implication of the Causally Interpreted Bell–Everett Theory

Abstract

The "many worlds" of Everett's can be reduced to temporal slices which are lined up on the time axis — to a shish kebab of worlds, as it were. This interpretation of quantum mechanics is equivalent to Everett's interpretation, Bell showed. Like in the original Everett theory, there exists a bundle of internally consistent worlds that are each complete with its own past and future. However, in addition a rapid hopping across the bundle occurs — with only one fiber "illuminated" at a time (real at that moment). This bizarre result from quantum mechanics turns out to be in accord with endophysical arguments.

The internal observer of an artificial molecular-dynamics-simulated universe sees only the "difference" between his own internal micromotions and the rest. Hence only one out of many possible "interface worlds" is valid for him or her at any given moment. It is governed by a Nelson type diffusion law (so that objects acquire quantum-mechanical properties). More important, perhaps, is the fact that the interface is time-dependent. It follows that the momentarily valid "eigen state" of a measured object changes rapidly - without this fact being accessible to the internal observer. Every single "eigen world" thus corresponds to a fiber in the Everett bundle.

The causal endophysical explanation offers the advantage that more can be said than is possible in the quantum analog. The hopping across fibers is again unavoidable. However, it at the same time depends on a parameter which can be indicated. Any change in the "assignment" of a single elementary particle (of whether it belongs to the observer or to the rest of the universe) is already sufficient to put a different fiber in charge. This means that the operator of the next-upper level of reality (in front of his exo-keyboard) controls an added free parameter when running the show for the internal observer. A slight displacement on the "piano" of assignment lets certain fibers be in charge more frequently. While it will be hard to consistently favor a single fiber, favoring a "thread" of adjacent fibers may be possible. The hopping then becomes restricted to a subbundle. Some such threads of mutually related worlds will be "consistent" in the sense that in the one thread, Schrödinger's cat always emerges unscathed while never in the other.

To a human observer sitting in front of a discrete state keyboard and lower level universe, a problem arises in the analogous case: picking a rare thread is most likely an NP-complete or even Gödel-impossible task. To circumvent this problem, a "Laplacian desk-top computer" would be needed. The latter — unlike present day computers — sports an inconspicuous switch labeled "turbo" (boosting the number of digits per word to infinity without any drop in speed). It might then as well possess a little tunable knob in addition, labeled "hue" (allowing the user to adjust the degree of living color or shadow-world-likeness, respectively, of the momentarily picked thread). The least sorrowful and most colorful thread is "paradise thread." On the Laplacian computer, the choice of the paradise option might turn out to be protected by a password. Depending on whether or not it is activated, the icon "FS" (for "flaming sword") or "OR" (for "Orpheus") would appear on-screen.

However, even if one reluctantly leaves the seat of the superobserver, the model universe approach remains a challenge. For the single exo world that exists behind the momentary thread can perhaps be "tapped" by the internal observer. "Blindsight experiments" permitting a certain degree of thread selection (and by implication world selection) are conceivable. Eventually, the internal observer may learn, for example, to manipulate his own curved hypersurface in exo space-time (into which his current fiber and thread are embedded) to a sufficient degree to bring it in contact with another thread belonging to another, not necessarily contemporaneous, observer ("greeting with the glove"). The iron bars of the world would then acquire spacings. Due to its enhanced nonlinearity, the momentary micro hypersurface lets time travel and the wish-machine-repair-shop appear somewhat less exotic today than they did at the time they were first proposed by Kurt Gödel and Gilles Deleuze (and Jonas[1]), respectively.

For J. O. R.

References

1. *Jonas' Welt — Das Denken eines Kindes* ["Jonas' World — The Thinking of a Child," R. Rössler and O. E. Rössler, eds.], pp. 77–78 (Reinbek, Rowohlt, 1994).

13

Into the Same Rivers We Step and Step Not, We Are the Same and We Are Not: On the Origin of the Now

Summary

Is time just a parameter in an otherwise unchanging universe, or is there "invisible" change in existence as Heraclitus claimed? According to a new interpretation of quantum physics, there is a different world in charge at every moment. The combined theory of Deutsch, Bell and Everett implies that one physical reality goes with one now and vice versa. While this interpretation of quantum mechanics is a minority opinion in quantum mechanics, the same structure unexpectedly arises in a classical model as an implication. The "interface" between a classical chaotic observer and the rest of his or her artificial universe possesses the same hidden time dependency. A new physical limit — impossibility of a direction-of-time clock — is implicit. Six "functions" can be attributed to the physical now so far.

13.1 Introduction

Heraclitus — undeservedly called "the obscure" — was fascinated by the phenomenon of change, as is well known. Unfortunately, only short fragments of what he wrote around 500 B.C.E. are extant. Aristotle gave a neat summary of what Heraclitus had wanted to say (for example, in his fragment no. 49a quoted in the above title) when he wrote: "[The Heracliteans] say that all things are in motion all the time, but that this escapes our perception" (Ref. 1, p. 197).

There is another obscure notion in existence that each of us uses every day without having the slightest idea of what it means: the little word "now." It will be proposed in the following that it was this even more difficult problem that Heraclitus grappled with.

At first sight, the notion of the now belongs to the humanities since "now-ness" is synonymous with "being conscious" and consciousness has no place in physics. Even worse, physics has actively removed itself from having a voice in this context. It destroyed the last bridge to conscious experience when it showed that objective simultaneity — a precondition for the existence of an objective now — cannot be uniquely defined and hence does not exist. This, at least, is Gödel's[2] interpretation of Einstein's and Minkowski's results.

There exists, however, one branch of physics — quantum mechanics — which is incompatible with relativity theory.[3,3a] It claims that the world is "created" by the observer at certain moments. For example, the "collapse of the wave function" (which is responsible for the occurrence, or not, of a "click" in a Geiger counter, for example) is believed not to take place unless some-one listens. "The tie between past and future has been cut," says Wheeler.[4] This is the so-called "measurement problem" of quantum mechanics which has never been solved. Like an open wound, it only waits for the pepper of the philosophers to be poured over it, so to speak.

Fortunately there exists a new development in physics — so that we may continue. At first sight, it belongs more to philosophy than to science because it relies on a certain "interpretation" of quantum mechanics — and interpre-tations are often considered cheap, because they have no "tangible" (testable) implications. However, Deutsch[5] arrived at a new implication — so that a falsifiable paradigm is in the making.

13.2   The Theory of Everett, Bell and Deutsch

Like Copernicus' new theory which was simpler but introduced vast amounts of "useless empty space," Everett's[6] theory invites the criticism of violating Occam's razor of maximum parsimony. The quantum decisions which occur in a measurement do not "really" take place, according to this theory. Rather, the whole wave front of potentialities described by the wave function of quantum mechanics exists as a set of actually realized events — only that the latter some-how are "shielded" from each other like the branches of a tree. The fact that to us invariably only one world shines in its actuality is explained by Everett by means of the word "relative." The unknown microstate of the observer selects the branch valid "relative to it" (namely, to this state). This is the meaning of the words "'relative-state' formulation" in the title of Everett's paper.[6] The disruption of the formalism by the usual reduction postulate ("collapse of the wave function") is thereby avoided.

The fact that no detailed account of the momentary microstate of the observer is given in Everett's theory has been noted (Hans Primas, personal communication, 1995), but represents no liability compared to the other, more familiar interpretations of quantum mechanics. This is because the price they have to pay for the unexplained reduction postulate — the postulate that a new wave function arises out of the blue — is even steeper from the point of view of Occam's razor.

Nevertheless only a minority of physicists — for the most part, cosmologists — "believe" in Everett's theory of world splitting. The reason they prefer it is that the observer no longer plays an active (state-creating) role. The free choice of the observer as to which subset — of a set of mutually incompatible observables — to turn into a reality by means of his or her measuring apparatus, is no longer crucial since world states exist relative to the observer anyhow (although not in the form of an eigen state). This is crucial under a condition in which there is no machine available into which to put the whole cosmos when aiming to measure its quantum state — as the competing "Copenhagen interpretation" seems to require.

In 1981, Bell[7] made a new contribution to Everett's theory. He showed that the usual understanding of Everett's theory — which assumes that all the different branches ("worlds") exist simultaneously — is not obligatory. A very rapid "hopping," from one world to the next, is equally in accord with the formalism of quantum mechanics. Even if for some reason more than one world should exist simultaneously in accord with the usual understanding of Everett's theory, each containing a clone of the observer, this would still be to no avail. For in this case, too, nothing guarantees that each clone is given the privilege of staying in one and the same world continually. The case of incessant "hopping" is again the more natural one to assume from the point of view of the formalism. Hence it indeed suffices to assume a single observer in a sequence of worlds.

This interpretation sounds absurd — unless one equates the "rivers" of Heraclitus with the "worlds" of the Bell–Everett theory. Then one darkness illuminates the other. The story does not end here, however. A third element has been added by Deutsch.[5] He recognized that, if (as shown by Everett[6]) there is, to each experiencing observer, only one world accessible "at a time," this fact is tantamount to a physical corollary existing to the now: a world.

It is all only wordplay. "At a time" has two meanings: the usual nontemporal one, and the literal one. Maybe arriving at this particular Koestlerian "bisociation" (to put together two previously unconnected facts in a creative flash) would have been impossible to achieve in any other language. At any rate, the newfound identity can immediately be exploited.

A first way to do so was already seen by Deutsch himself. He proposed a new class of experiments which, when carried out by a conscious observer, allow him or her to obtain new results about the physical world. (Specifically, he proposed looking for residual information from a previous Everett world.) The empirical fact of nowness is thereby introduced as a new sentinel into physics. This is only one possible way to exploit the new identity between nowness and the world. In the present context, the interest lies in the other direction of the identity. Is it possible to, by using physics, describe the origin of the now?

13.3   Nowness as a Gödel Boundary

The proposed second option to exploit Deutsch's identity between now and the quantum world immediately runs into a difficulty. The phenomenon of state splitting between observer and rest is not rigorously explained in Everett's formalism since the new physical correlate to nowness — the observer state before the rest of the universe manifests a state "relative to this state" — remains unspecified. Hence nowness remains unspecified in physical terms, too. To arrive at the desired explanation of nowness, first the whole of quantum mechanics (including state reduction) would have to be explained physically.

Fortunately, there exists a shortcut. Since it is "only" nowness which is at stake, it is not the whole of physics that needs to be explained first. It suffices to set up a little *model* which shares sufficiently many features with Everett's theory to pass for a caricature. An "analog to nowness" may then still be implied — in the model. And more than an analog cannot be hoped for anyhow in science.

The great "eye-opener" in modern physics is Gödel's[8] mathematical discovery that there exist truths which are not accessible in a finite number of steps from within the formal system which implies them. The decisive notion is "from within." Gödel would never have reached his result if stepping out of the formalism had not been an option. The fundamental insight thereby gained is: "Endophysics is different from exophysics."[9] That is, physical universes, too, are bound to look different when observed from the inside than when observed from the outside. An example for a limit to self-observation of a finite automaton was given by Finkelstein and Finkelstein[10] (cf. Svozil,[10a] who also mentions earlier work by Moore[10b] and Toffoli.[10c]).

The notion of a "Gödel boundary" (between accessible-from-the-inside and inaccessible-from-the-inside truths) is due to Hofstadter[11] (see Fig. 18 of his book[11]). The idea that specifically the quantum-mechanical indeterminacy might be reducible to a Gödel type mechanism was first uttered in print by Gödel's close friend John von Neumann: The "result of the measurement" might be "indeterminate" because "the state of information of the observer regarding his own state could have absolute limitations, by the laws of nature"[12] (cf. Chap. 6). Later Popper, who had paid Gödel a visit, expressed a similar view in a classical context.[13] Gödel himself in his late years confided to John Wheeler — after the latter had for the first time successfully steered the topic of conversation into quantum mechanics: "I just don't believe in it" (John Wheeler, personal communication, 1983). Gödel apparently also reckoned with an ultimate explanation — in Gödelian terms.

Nevertheless a Gödelian theory of quantum mechanics or physics has, apparently, never been put forward. Two reasons can be given. (1) The original Gödel formalism would have to be generalized, away from the realm of discrete arithmetic, to the calculator of continuous dynamical systems (Roland Wais, personal communication, 1981). (2) The physical ideas of both von Neumann and Popper, while inspired by Gödel, involve a complication absent in Gödel's theory: they call for an explicit inclusion of the observer in the formalism. The Gödel boundary would then have to be described in relation to an observer. This new constraint is, by the way, independent of whether the artificial universe is continuous or discrete.

At present, to generate an explicit observer through microscopic simulation — as a recurrent "dissipative structure" in the sense of Prigogine; cf. Ref. 13a — is still beyond the reach of current computing power. This applies both to continuous and to discrete models. (Ultimately, all computer models remain discrete.) What really has to be accomplished, however, is an analysis of the "interface" between the internal observer and the rest of the model universe. To define the interface numerically — so that it can be put on-screen — defies current methods to investigate the qualitative behavior of dynamical systems.[13b] It is already evident that, once the problem has been solved, a giant computational "overhead" will be implicit compared to a "faceless" calculation of exactly the same universe. Moreover, if the universe in question is more than one-dimensional, an "NP-complete" computational step may have to be reckoned with which would thwart the numerical realizability of a "realistic" observer interface.

Thus, much as in the case of Everett's theory, properties of the interface are hard to derive even under classical conditions. The present proposal therefore may have little chance to succeed soon — unless one stumbles across a specific feature of an explicit classical interface which on the one hand is Gödelian in type and on the other is derivable "the short way." A candidate for such a property exists.

13.4   Campbell's Idea Concerning the Structure of Time

In 1927, Norman Campbell[14] reproposed an idea he had already sketched six years earlier.[15] All quantum phenomena might be reducible to a single unexplained axiom: "Time is like temperature." What he had in mind was that time may be well defined only for objects of high mass (large particle number) while for low mass objects, an undefinedness should set in analogous to that of temperature which ceases to be well defined for low particle numbers, too. This postulate, which inspired Bohr to his well-known complementarity postulate,[15a] went into oblivion.

Whether Imre Fényes took it up may still be worth checking. At any rate Nelson[16] (who quotes Fényes[17]) arrived at another postulate several decades later that is related in a sense. According to the single axiom of "stochastic mechanics,"[16] all objects of mass M are subject to an unexplained "diffusion coefficient" D of size h/M, where h is Planck's constant divided by 4π. Unexpectedly, many properties of quantum mechanics — including validity of the Schrödinger equation — are a mathematical implication of this axiom.[16] Yet, what could be the reason for this axiom to hold good in the real world?

There exists a "third axiom," related to Campbell's, that implies Nelson's. It is, moreover, reducible to a Gödel type argument. It posits that the *direction of time* is not well defined on a micro scale. More specifically, it consists of the claim that a "direction-of-time clock" cannot be built or used when very short periods of time are at stake.

This new axiom is a bit in the spirit of Einstein's discovery that simultaneity is not something to be taken for granted but something to be measured. If the axiom is correct, a new theory in the spirit of relativity, but applicable to the micro realm, is implicit. It turns out that quantum mechanics can be formulated in such a way that the new axiom is an implication. Bialynicky-Birula's[18] and Zambrini's[19] results allow this interpretation. Since formal implications of quantum mechanics are as good as facts of nature (with no exception so far), this fact may be taken as evidence that the new axiom is indeed valid in the real world.

The new empirical postulate can then be used to "explain" quantum mechanics — via the Nelson formalism. (This feat has apparently already been accomplished by Nagasawa.[19a,19b]) However, the story is not finished here. On the one hand, the reason why the new axiom holds good in nature remains unaddressed. On the other, stochastic mechanics is still beset with difficulties (the problem of nonlocality is still not completely solved[20,19b]). Thirdly, and most important in the present context, the problem of nowness has not come any closer to a physical understanding.

13.5 The Return of the Observer

Given the fact that the new axiom ("microvacillation of time's axis") is admissible, the question of whether or not it can be derived in a Gödelian fashion arises. The hope is that, simultaneously, some light will be thrown on the problem of nowness.

Two general possibilities are open, the first "general Gödelian" in its type, the other "special Gödelian." The former would consist of the demonstration that the new axiom about time possesses general validity in the model universe in question. That is, the causality vacillation would be a property not of the interface between one particular explicit internal observer and the rest of the universe but of all interfaces — as a "shared property" valid in a public interface that comprises all observers that exist or are possible. The "quantum mechanics" implicit in this axiom would again represent a "maximally objective" theory. Although interface-bound, it would muster the utmost generality that any theory valid from the interior of a universe can hope to possess. The program of an "objective physics" would be vindicated once more.

The second possibility is both more modest and more far-reaching. Again, one would arrive at an "interface-bound" explanation. However, the degree of generality of this explanation would be diminished. While for an individual internal observer, the interface would still be all he has (so that it would be universal to him), it would cease to be truly universal. Part of the "objective" (to him) properties of the world would not be objective at all. He would be a prisoner, not only of his own universe (as everyone is) but also of his own brain. While this holds good also on the macro level, it leaves the world itself unaffected (cf. Chap. 15). In the micro case envisaged here, however, objective physics itself would become a part of the "nonobjective half" of nature. For the Gödel boundary would cease to be objective in a "broad" sense (like a canopy), becoming "narrowly fitting" (like a swimsuit).

It almost goes without saying that the second alternative actually is the more general one. It would be naive to hope that once the problem of how the interface looks like for a single observer has been solved, the properties of that interface will carry over literally to the next observer. Any Gödel type approach based on internal observers has a solipsistic streak to it. The new paradigm of an "observer-dependent Gödel boundary" would thus, if confirmed, bring along an even more radical change in the understanding of the world than this was achieved already by the "anonymous" observer of Bohr's quantum mechanics.

13.6   Effectively Quasiperiodic Observers

There is one class of classical observers for which the temporal impossibility axiom of Sec. 13.4 (absence of a consistent direction of time) is an implication. It is the class of "effectively quasiperiodic observers."

Hyperchaotic classical multiparticle systems in one dimension are "effectively quasiperiodic."[21] That is, almost all Lyapunov characteristic exponents (which indicate the exponential divergence of neighboring trajectories) are effectively zero.[21] The motion in phase space (a moving point in $2N$-dimensional space which describes the behavior of the N-particle system in 1-D real space) is, therefore, essentially confined to a (multiple-connected; cf. Ref. 22) "torus." A torus is the surface of a doughnut, and the motion on it resembles the never-crossing winding of an endless thread wound around a wooden doughnut. A special feature of the present hypertorus is that all its circumferences are statistically equal — since the different particles in their individual one-dimensional classical Pauli cells in real space[21] all have the same mean period.

As a consequence, every motion comes in two varieties, one running forward in time and the other backward in time. That is, we are essentially faced with a collection of equal pendulums, each first going forward, then backward, and so on. Hence most momentary motions recur in a similar fashion after a mean half period, having their momenta reversed. Therefore "effective time reversals" occur in rapid succession.

This "quasiperiodization" result[21] is robust because it applies also to macroscopic dissipative structures generated by the hyperchaotic micro dynamics — internal observers included. The macroscopic time's arrow valid for the dissipative structure in question therefore paradoxically does not imply the existence of a "micro arrow" that would run parallel to the macro arrow. Rather, the observer consists of "time slices" belonging to two equivalence classes that are equivalent under time reversal.

Nevertheless it would be misleading to say that half of the time slices run "parallel" to the macro arrow and the other half "antiparallel." The two types of microscopic motion contribute to the macro arrow on an equal footing. Nevertheless it is correct to say that the observer is effectively "chopped up" into short segments such that almost any trajectorial segment is virtually identical to a nearby one except for its orientation in time. Thus the observer consists of two "subobservers" in effect, one facing one way in time and the other the other way. Both have been chopped up and intermingled: a "black" (or dissipative) subobserver and a "white" (or antidissipative) subobserver.

A different way to put the same result is to say that the observer — like any other dissipative structure based on indistinguishable particles — is "effectively at equilibrium." For at equilibrium, such a "mosaic type" structure (forward and backward) would not be surprising. The fact that "effectively quasiperiodic" systems are "effectively at equilibrium"[9] would thus be the real explanation of an otherwise paradoxical result.

13.7 Classical Clocks' Limit

We now come to the question of observer-external clocks in the classical model universe. It goes without saying that to an observer of the causally vacillating type, the orientation in time of external motions becomes a problem. He would need a "clock" telling him, with respect to a given external particle he observes, "what direction of time it is" for that particle, momentarily.

The same question poses no problem, of course, in the exo world of the superobserver who has the universe in residence in his computer. For the superobserver it is easy to relate all events in the simulated world to the "objective direction of time" which results from Boltzmann's arrow in the computed universe. (This arrow, by the way, is independent of whether the universe in question is run backward or forward in the computer, as is indeed possible to do in both directions in an identical fashion.[22a,b]) The objective macro direction of time is deterministically defined — by the systematic increase of the momentary phase space volume.[13a] A method of calculating this deterministic entropy in a molecular dynamics simulation is already available (Hans Diebner, personal communication, 1996; cf. Ref. 22c).

Our question had been a different one, however: Can the momentarily valid directional information (on the relationship between the macro arrow and a micro motion) be made available to the internal observer? So that he would know the color (whether "black" or "white") of his own momentaray trajectorial segment?

This task is indeed insoluble for the internal observer. For he cannot step outside of himself in order to "read off" the clock and then go back inside to make use of that reading. Rather, he has to observe the "clock" on the same footing as he does any other external particle. Since the "hands" of that clock need to perform very fast motions indeed, the hands unavoidably have to be represented by small fast-moving particles themselves in that mechanical universe.

We therefore encounter a new Gödel type situation: an object's direction of motion can only be determined through comparison with a standard motion. Since any standard motion must be the motion of an object particle, too, an infinite regress applies. As long as the observer cannot step out of himself, no way out appears to be in sight. Only an external observer — observing the observer — could solve the problem for the first observer. He might, for example, effectively shut off the sensors of the first observer during every second time slice valid for the latter. In order to be able to do this, however, the external observer would either have to be completely outside the universe of the first (having it in his own computer) or have to possess time slices that are much more fine-grained than those of the observer to be observed.

The last alternative breaks down in a chaotic universe. It is then in principle impossible to ascertain, through interaction with many individual chaotic particle trajectories of the observer over a sufficiently long time, the beginning and the end of the next time slice valid for the observer. Chaos theory — specifically its "sensitive dependence on initial conditions" (as David Ruelle put it) — rules out a state of sufficient information to be achievable by the observer's observer. Therefore only the regressus-free higher level solution remains open; from the inside of the universe in question, the new temporal uncertainty appears to be insurmountable.

Hence a direction-of-time clock is indeed impossible on the finest scale. Boltzmann's analogous result in the macro domain — which showed that the time's arrow of the cosmos cannot be determined objectively from within the cosmos — acquires a microscopic analog.

13.8   The Problem of "Smart Machines"

Before we can make the connection with the problem of the now, we still need to better understand the model universe. What consequences does the clocks' limit have for the observable behavior of a micro particle? Does the latter — for example — perhaps appear "blurred"?

At first it makes sense to ask whether the new temporal limitation cannot be "outwitted" by the use of smart machines. For, as in thermodynamics, it is not easy to believe that something that can be done from the outside (by one's "playing demon") should be impossible to achieve from the inside. How does an external particle appear to an observer who makes use of a measuring chain of the amplifying type?

The observer is the sum total of the dynamical processes which go on inside him or her. There is a macro dynamics (the coarse-grained responses of the dissipative structure called "observer"), and there is an underlying, much faster microdynamics. Even the most rapid macro change in the observer lasts several orders of magnitude longer than a micro time slice does. The micro time slices therefore are necessarily "integrated over" from the macro point of view. An analog to "flicker fusion" — but much faster — can thus be predicted to hold good for micro time slices.

As far as macroscopic external objects are concerned, this prediction is correct. However, when the observation of a micro object's motion is at stake, use of an amplifying measuring chain (sporting a macro pointer at its end) is required in the mechanical universe. This pointer provides a both smoothed-out and delayed momentary record of what the micro particle at the front end of the measuring chain does or just did. Since the pointer is an ordinary macro object, one expects that no fundamental limitation to the accuracy of measurement should make itself felt, even if the behavior of a micro object is to be ascertained with the aid of the pointer.

This expectation is valid on the exo level. However, it is at variance with the above limitation (that an objective measurement of a micro motion including its temporal orientation is impossible from the inside). Therefore something appears to be missing in the statement made above. Highly smoothed-out and hence inaccurate recordings of a particle's position are only one type of measurement the internal observer is interested in. The second type is a highly resolved representation of the particle's motion. Hereby an accurate localization in both space and time is essential. Different pointers connected to stationary sensors could be used to monitor the particle's path objectively. Also, the result need not be registered immediately. It could be recorded and later played back in slow motion or be used for controlling a plotting device, for example. Thus once more no trace of the previous dilemma makes itself felt.

However, that would mean that the observer can have the micro world presented to him in an objective fashion by means of appropriate smart machinery. The impossibility of a direction-of-time clock would be effectively circumvented. Where does the error lie? Has perhaps the principle that the observer cannot step out of himself been violated in the above scenario? To see that this is indeed the case, a closer look at the behavior of the measuring chain is justified.

13.9   The Prediction of Blurredness

Any measuring chain which presents a micro object to the observer in a certain way, during one time slice, continues to do so in the next time slice but with the opposite temporal orientation (with jumps occurring at the connection). The passive, amplifying, causal chain of the first time slice becomes an active, disamplifying, anticausal chain in the second time slice. The integral taken over both time slices is therefore effectively equivalent to the integral that would be valid under a reversible coupling between observer and object.[23]

It follows that the internal thermal noise of the observer, present while the observer mechanically absorbs the pressure exerted by the pointer, acts not only on the pointer. Rather, it propagates through the whole measuring chain up to the external particle. The particle "responds" by making essentially the opposite motion compared to what it did during the preceding time slice. Thus, an energetic perturbation is seemingly absorbed by the particle every second time slice.

In this way an "effective diffusion coefficient" arises which is independent of the amplification provided by the measuring chain. It has the form ET/M. That is, it depends on the length of a time slice, T, on the thermal noise energy of the observer, E, and on the object's mass, M. Essentially the same formula was encountered above (Sec. 13.4). Thus a connection between stochastic mechanics (a version of quantum mechanics) and the present clocks' theorem appears to exist. The close relation between the "no-micro-clock property" of quantum mechanics and its diffusive properties, mentioned above, has found a classical analog.

At the same time a major discrepancy becomes visible between the two theories: the object's diffusion does not stop after the measurement in the classical analog. For if the diffusion stopped, after a measurement, the "clock" would suddenly acquire a well-defined reading for a distinct past moment in time. Even this appears to be blocked in the classical case.

The model universe therefore predicts a "quantum mechanics more quantum-mechanical than quantum mechanics." A set of highly resolving pointers would indeed present the object in a "blur" to the internal observer. Or, to make the consequences more vivid, a "prey" (like a butterfly) living in the model universe could in the course of evolution learn to "camouflage" its own location by letting its flight trajectory be controlled by the amplified behavior of an internal micro particle. As a "living pointer," it would get blurred to the point of being invisible.

13.10   Nowness-vs.-Blurredness Trade-off

We saw that the "built-in" reduction of the wave function, once a measurement has occurred in stochastic mechanics,[16] is lacking in the model universe. It follows that the "hull function" — defined by the linear wave equation of quantum mechanics — remains in charge continually for a particle even after it has already been "measured." Measurement thereby ceases to be a finished business. All measuring chains remain in charge continually while growing in their temporal lengths. Thus a "quantum world" of the type originally envisioned by Schrödinger appears to be in charge in the model universe at first sight.

This conclusion hinges on the assumption that the integration over the different time slices continues indefinitely. However, there necessarily exists a limit to the integration interval since the observer changes his macro state on a macro time scale. Accordingly, only a finite — flicker-fusion-like — time interval appears to be available for the integration. The blurredness is therefore confined to that interval. This means that after every "macro time slice," a new "quantum world" of again restricted blurredness is in charge. Moreover, portions of the wave function that are characterized by sufficiently small "amplitudes" are no longer represented, during the macro time slice. This means that a kind of "partial reduction" of the wave function occurs during every macro time slice — that is, during every "now."

In this way, a function of the now suddenly shows up in the present model universe: the now partially reduces the blurredness of the wave function.

13.11   Functions of the Now

If we assume that for some reason the integration interval were reduced more and more, the "partially reduced" world of a long integration interval would become more and more "thoroughly reduced." Hereby larger and larger portions of the wave function (up to fairly large amplitudes eventually) are cut out in effect. In the limit — as the integration interval becomes very short — "complete reduction" is obtained. Only in cases where no measurement is made does the superposition of worlds remain in charge. The observed "sustained interference" would be a reminder of what held good before the integration interval was reduced.

What is the price to pay for the "nows" of decreasing lengths, introduced above? There are two punishments: nonuniqueness and hermeticity. The uniqueness disappears at the very moment the integration interval ceases to be infinitely long. For in the next (very long) now, a different portion of the low amplitude parts of the wave function will be missing, as we saw. On the other hand, it is correct to say that the low-amplitude parts that are removed affect the overall picture only to a very minor degree when being absent so that the violation of uniqueness is mild. The fact that in the next macro time slice other parts of the wave function figure prominently through their being absent will be virtually undetectable in the common blur that remains.

Eventually, however, as the duration of the integration interval ("now") is made shorter and shorter, the successive "now worlds" differ in more and more details. At the end, a wild "hopping" occurs, just as in the Everett case.

We thereby have arrived at the second implication — hermeticity. Since the blur is an artifact caused by the superposition of micro-time-reversal-generated effects, it does not stop short of involving memory. Hence when a different blur is in charge in the next "now," the old now's blur and all traces of it are forgotten. This hermeticity remains valid as the integration intervals (nows) become shorter and shorter in duration. Hence a kind of "temporal trap" restricts the observer to a now world from which there is no escape. For no matter what its duration, each now world comes equipped with its own past and its own future. Thus a picture very close to that depicted by Deutsch, Bell and Everett (cf. Sec. 13.2) emerges.

The notion of the "integration interval" thus constitutes a tunable parameter with the aid of which it becomes possible to better understand some of the strange features of nowness — as they are valid in the model universe.

Summing up, six "functions" of the now could be identified so far in this manner:

(1) If the integration interval is unbounded, the corresponding "now" of infinite duration ensures perfect superposition of all alternative states of micro objects and clocks presenting themselves to the observer. That is, the now preserves the falsity of the superposition compared to the exo truth.

(2) If the integration interval is finite, every now "reduces" the superposition in its sufficiently low amplitude parts. That is, the now destroys the uniqueness of the full superposition.

(3) The now thereby again generates a "false" picture — both from the viewpoint of the unreduced superposition and from the viewpoint of alternative reduced superpositions generated by other nows.

(4) The now protects the "false uniqueness" from being detectable.

(5) The now prevents the whole chain of "hidden change" from being detectable.

(6) In the limit of arbitrarily short finite integration intervals, the nonreduced (nonmeasured) portions of the wave function retain their influence by exhibiting "interference."

If the theory were applicable to our own world, a reason would have to be found as to why only "maximally short nows" are in charge in the real world.

13.12 Conclusions

To think that the philosophical problem of the "now" may be accessible to scientific investigation appears naive at first sight. Exempt from this verdict would be physiological investigations into the dynamics of the short term representation of the world in the so-called working memory of the brain. The brain is hereby looked at as an "ordinary" (macroscopic) machine.[24]

However, just as there exists a "physiological nowness window," there apparently also exists a "physical nowness window." Deutsch[5] realized that a now world is an Everett world. This insight implies that in time, a kind of "tunnel vision" exists. Whole worlds exist for the observer — one at a time — for a very short time but such that each contains its own consistent past and future. The "tunnel" acquires a "tail," so to speak. Every moment thereby becomes a death and a birth into a new world.

While compatible with modern physics,[6,7] this view does not follow from it as an inescapable conclusion since alternative interpretations like the standard ("Copenhagen") interpretation remain possible. Therefore it is a challenge to try and reproduce the same state of affairs under conditions in which the successive-worlds interpretation is the only one possible. In this way a "model" of the origin of the now would become accessible. A preliminary attempt in this direction has been presented.

Most scientific results are trivial from a philosophical point of view. Even if nowness — in the sense of a "nail" which cuts out free traffic along the time axis — could be explained, the origin of the subjective now would remain shrouded in mystery. Only if one already takes for granted that a subjective correlate to certain physical states exists, is it of interest to arrive at a "narrowed-down" class of candidate states. The "states" defined by a short integration window in the model world share with the subjective now the formal property that a whole world is attached to each in such a way that neither of the two exists without the other.

The approach includes an observer in a reversible model universe. Bohr's paradigm of a "generalized observer" — the scientific community conversing in a natural language on a classical macro level — is replaced by an individual observer in the model. "Solipsism" therefore seemingly rears its head. A Gödel boundary between an observer and the rest of the universe is almost never a public affair. It is this insight which distinguishes the present model physics from traditional physics.

Could it be that the time has come to abandon "our beautiful physics" in order to enter the no longer observer-independent "second phase" of its history?

The new notion of the interface implicitly invokes consciousness, although only indirectly. If consciousness exists in the physical world, it apparently cannot do this without being attached to an interface. An interface is intangible — a cut of zero measure without any substance. It is surprising that a theory appears to be possible. Quantum mechanics, on the other hand, has long grown accustomed to the idea of invoking consciousness as the ultimate medium in which reality materializes, so to speak. For his courage to do so — in continuation of work by his late friend John von Neumann — the present paper is dedicated to Eugene Wigner.[31] However, there exists an even older philosopher–scientist to whom the present situation would have looked familiar — the "obscure" Ephesian mentioned in the Introduction. "To those who step into the same rivers, different and different waters flow — and (different and different) souls emerge from the humid" (Fragment No. 12,[1] p. 196).

Acknowledgments

Chapter originally presented at the 16th Institute of Pedagogics in Science (IPN) Workshop, "Nonlinear Processes and Natural-Science Education," held March 14 18, 1988, in Kiel. I thank Jan Robert Bloch and his group for stimulation. I also thank John Nicolis, Hans Primas, Michael Thorwart and Mohamed El Naschie for discussions.

References

1. G. S. Kirk and J. E. Raven, *The Presocratic Philosophers — A Critical History with a Selection of Texts* (Cambridge: At the University Press, 1957), pp. 198, 197, 196.

2. K. Gödel, "A remark on the relations between relativity theory and idealistic philosophy," in *Albert Einstein — Philosopher Scientist* (P. A. Schilpp, ed.), pp. 555–562 (LaSalle: Open Court, 1949).

3. A. Einstein, B. Podolsky and N. Rosen, "Can quantum-mechanical description of physical reality be considered complete?", *Phys. Rev.* **47**, 777–780 (1935).

3a. L. Landau and R. Peierls, "Extension of the indeterminacy principle for relativistic quantum theory" (in German), *Z. Phys.* **69**, 56–68 (1931).

4. J. A. Wheeler, "The experiments of delayed decision and the dialogue between Bohr and Einstein" (in German), in *Moderne Naturphilosophie* (B. Kanitscheider, ed.), pp. 203–222 (Würzburg: Königshausen und Neumann, 1984); p. 205.

5. D. Deutsch, "Three connections between Everett's interpretation and experiment," in *Quantum Concepts in Space and Time* (R. Penrose and C. J. Isham, eds.), pp. 215–225 (Oxford: Clarendon, 1986).

6. H. Everett III, "'Relative-state' formulation of quantum mechanics," *Rev. Mod. Phys.* **29**, 454–462 (1957); reprinted in *The Many-Worlds Interpretation of Quantum Mechanics* (B. S. DeWitt and N. Graham, eds.) (Princeton: Princeton University Press, 1973), pp. 141–149.

7. J. S. Bell, "Quantum mechanics for cosmologists," in *Quantum Gravity 2* (C. Isham, R. Penrose and D. Sciama, eds.), pp. 611–637 (Oxford: Clarendon, 1981); reprinted in his *Speakable and Unspeakable in Quantum Mechanics* (Cambridge: Cambridge University Press, 1987), pp. 117–138.

8. K. Gödel, *On Formally Undecidable Propositions* (New York: Basic Books, 1962); German originally in *Monatshefte für Mathematik und Physik* **38**, 173–198 (1931); see also: W. DePauli-Schimanovich and P. Weibel, *Kurt Gödel-Ein mathematischer Mythos* ["Kurt Gödel — A Mathematical Saga"] (hpt-Verlag, Vienna, 1997).

9. O. E. Rössler, Endophysics, in *Real Brains, Artificial Minds* (J. L. Casti and A. Karlqvist, eds.), pp. 25–46 (New York: North-Holland, 1987); cf. also Chap. 6 of the present book.

10. D. Finkelstein and S. R. Finkelstein, "Computer interactivity simulates quantum complementarity," *Int. J. Theor. Phys.* **22**, 753–779 (1983).

10a. K. Svozil, *Randomness and Undecidability in Physics* (Singapore: World Scientific, 1993).

10b. E. F. Moore, "Gedanken-experiments on sequential machines," in *Automata Studies* (C. E. S. McCarthy and J. McCarthy, eds.) (Princeton: Princeton University Press, 1956).

10c. T. Toffoli, "The role of the observer in uniform systems," in *Applied General Systems Research* (G. Klir, ed.) (Plenum, New York, 1978).

11. D. R. Hofstadter, *Gödel, Escher, Bach — An Eternal Golden Braid* (New York: Basic Books, 1979), p. 71.

12. J. Von Neumann, *The Mathematical Foundations of Quantum Mechanics* (Princeton: Princeton University Press, 1955), p. 438. (German original 1932, Springer-Verlag, p. 233.)

13. K. R. Popper, "Indeterminism in quantum physics and in classical physics I," *Brit. J. Phil. Sci.* **1**, 117–133 (1951), p. 129.

13a. O. E. Rössler, "Explicit dissipative structures," in *Invited Papers Dedicated to Ilya Prigogine on the Occasion of His 70th Birthday* (A. van der Merwe and G. Nicolis, eds.), *Found. Phys.* **17**, 679–688 (1987).

13b. O. E. Rössler, "Four open problems in four dimensions," in *A Chaotic Hierarchy* (G. Baier and M. Klein, eds.), pp. 365–369 (Singapore: World Scientific, 1991).

14. N. Campbell, *Nature* **119**, 779 (1927).

15. N. Campbell, "Atomic structure," *Nature (London)* **107**, 170 (1921).

15a. H. J. Folse, *The Philosophy of Niels Bohr — The Framework of Complementarity* (Amsterdam: North-Holland, 1985), p. 105.

16. E. Nelson, "Derivation of the Schrödinger equation from Newtonian mechanics," *Phys. Rev.* **150**, 1079–1085 (1966).

17. I. Fényes, "A probability-theoretical explanation and interpretation of quantum mechanics" (in German), *Z. Phys.* **132**, 81–106 (1952).

18. I. Bialynicki-Birula, "Transition amplitudes versus transition probabilities and a reduplication of space–time," in *Quantum Concepts in Space and Time* (R. Penrose and C. J. Isham, eds.), pp. 226–235 (Oxford: Clarendon, 1986); I. Bialynicki-Birula, "Does measurement reverse the direction of intrinsic time?", in *Matter Wave Interferometry* (G. Badurek, H. Rauch and A. Zeilinger, eds.) (Amsterdam: North-Holland, 1988); *Physica* **B151**, 302–305 (1988).

19. J. C. Zambrini, "Probability in quantum mechanics according to E. Schrödinger," *Physica* **B151**, 327–331 (1988).

19a. N. Nagasawa, "Time reversal and Markov processes," *Nagoya Math. J.* **24**, 177–204 (1964).

19b. N. Nagasawa, "Quantum theory, theory of Brownian motions and relativity theory," in *Chaos, Information and Diffusion in Quantum Physics* (M. S. El Naschie, O. E. Rössler and G. Ord, eds.), *Chaos, Solitons & Fractals* **7**, 631–643 (1996).

20. E. Nelson, *Quantum Fluctuations* (Princeton: Princeton University Press, 1985).

21. O. E. Rössler and M. Hoffmann, "Quasiperiodization in classical hyperchaos," *J. Comp. Chem.* **8**, 510–515 (1987).

22. B. Eckhardt, J. Ford and F. Vivaldi, "Analytically solvable dynamical systems which are not integrable," *Physica* **D13**, 339 (1984).

22a. H. H. Diebner, "Investigations of exactly reversible algorithms for dynamics-simulations" (in German). Diploma Thesis in Physics (University of Tubingen, 1993).

22b. D. Levesque and L. Verlet, "Molecular dynamics and time reversibility," *J. Stat. Phys.* **72**, 519 (1993).

22c. H. H. Diebner and O. E. Rössler, "A deterministic entropy based on the instantaneous phase space volume," *Z. Naturforsch. a*, submitted.

23. O. E. Rössler, "A possible explanation of quantum mechanics," in *Advances in Information Systems Research, Design and Implementation of Information Systems* (G. E. Lasker, T. Koizumi and J. Pohl, eds.), pp. 581–589 (Windsor, Ontario: The International Institute for Advanced Studies in Systems Research and Cybernetics, 1991).

24. E. Pöppel, *Die Grenzen des Bewußtseins* ("The Boundaries of Consciousness") (Stuttgart: Deutsche Verlagsanstalt, 1985).

25. N. Bohr, "The quantum postulate and the recent development of atomic theory," *Nature* **121**, 580–590 (1928).

26. W. Heisenberg, "*Die Veränderungen des Wirklichkeitsbegriffs der exakten Naturwissenschaft*" ("The changed notion of reality in the exact sciences"), in *Die Mittwochsgesellschaft* (K. Scholder, ed.), pp. 332–333 (Berlin: Quadriga Verlagsbuchhandlung Severin und Siedler, 1982).

27. J. A. Wheeler and W. H. Zurek (eds.), *Quantum Theory of Measurement* (Princeton: Princeton University Press, 1983).

28. J. S. Bell, "On the Einstein–Podolsky Rosen paradox," *Physics* 1, 195–200 (1964); reprinted in his *Speakable and Unspeakable in Quantum Mechanics* (Cambridge: Cambridge University Press, 1987), pp. 14–21.

29. H. Primas, *Chemistry, Quantum Mechanics and Reductionism* (Berlin: Springer-Verlag, 1981).

30. O. E. Rössler, "Explicit observers," in *Optimum Structures in Heterogeneous Reaction Systems* (P. J. Plath, ed.), pp. 123–138 (Berlin: Springer-Verlag, 1989); cf. Chap. 8 of this book.

31. E. P. Wigner, "Remarks on the mind–body question," in *The Scientist Speculates* (I. J. Good, ed.), pp. 284–302 (London: William Heinemann, 1961).

14
A Possible Explanation of Spin

Abstract

Spin is one of the most elusive concepts in physics. It is a property of otherwise indistinguishable particles. Indistinguishability is a classical concept. It implies the existence of a "classical Pauli cell." However, the classical Pauli cell — unlike the quantum Pauli cell — contains not two but only one particle, for classical particles possess no spin, as Pauli stressed. Nevertheless spin arises endophysically — as a property of the "interface" which forms between an observing subsystem and the rest of its Hamiltonian universe, provided the latter is three-dimensional.

In a spatially one-dimensional universe, the "micro-time-reversals" in the observer only generate a back-and-forth perturbation in the environment. If the observer is two-dimensional, the micro-time-reversals no longer produce straight perturbations alone. The observer rotates in successive time slices in opposite directions relative to an external particle. The particle therefore performs the opposite rotation, relative to the observer, in the next, time-inverted ("white"), time slice. Nevertheless, the particle seemingly performs the same rotation as in the preceding black (non-time-inverted) time slice, as far as the observer is concerned. For during a white time slice, all external motions are perceived to take place in the opposite direction, as mentioned. Thus all particles always have the same spin in a two-dimensional universe.

Two spins arise in three dimensions. A spinning particle now in addition either approaches or recedes from the observer. This makes two otherwise equal particles distinguishable. Since, of particles in random motion, one half approaches and the other half recedes from the observer, they all fall into two different spin classes. More important, two equal particles in adjacent classical Pauli cells tend to have opposite spins. This makes them distinguishable — so that the two now share the combined cell. However, something strange happens during the next time slice: the partner which formerly had the one spin acquires the other and vice versa — the two swap their identities. This is because the direction of external path motions is inverted from one time slice to the next, as mentioned. The two particles become maximally correlated.

"Endo" Pauli cells having two correlated occupants therefore exist classically. If the same fact held good in the real world, "most" properties of atoms and molecules (Kossel pairs) and superfluids (Cooper pairs) would be nonexistent exophysically. It would be correct to say: Endochemistry is different from exochemistry.

Acknowledgments

I thank Peter Strasser, George Gokis and Niels Birbaumer for discussions. Michael Conrad kindly informed me, after receiving a copy of the present abstract, that he had independently arrived at similar conclusions. For J. O. R.

15
Microconstructivism

Summary

The perhaps most important lesson of brain theory is constructivism: reality is a pseudoreality because the "intuitive forms of perception" implicit in the dynamics of the brain constrain reality's properties. A cold lizard sees the world differently than a warm lizard does. Radical constructivism has so far been confined to the macroscopic realm. A "microscopic" version is proposed here. In it, the reversible micro dynamics which underlies the irreversible macro dynamics of the brain is focused on. For simplicity, a classical kinetic universe is presupposed. As in macro constructivism, an "interface" is formed which depends both on the observer-external dynamics and on the dynamics that goes on within the observer. Owing to the microscopic nature of the dynamics in question, the effects it exerts on the interface are by definition incorrigible by macroscopic means. That is, the effects cannot be recognized under ordinary circumstances by the observer in question. Specifically, the micro dynamics in the observer is subject to "effective time reversals" on a very short time scale. This is tantamount to the existence of a rapid "vacillation" of external causality. Hence the spatiotemporal properties of the environment are distorted in the interface in a highly nonlinear fashion. The predictions made by microconstructivism (or, synonymously, endophysics) include known and unknown features. Two features familiar from quantum mechanics fall in the first class (a unit-action noise and nonlocality). The unexpected predictions include a "double" nonlocality. The main implication — existence of an "observer-relative objective reality" — is, therefore, falsifiable.

15.1 Introduction

The significance of nonlinear dynamics for understanding the brain was pointed out by Zeeman,[1] Gurel[1a] and Tsuda.[2] Dynamical thinking is widely accepted today — after the discovery of the 40-Hertz rhythm of the brain,[3] for example. A direct connection with older "Darwinian" approaches to brain function[4] can be made.

If one accepts the legitimacy of a combined dynamical and evolutionary approach to brain function, precarious epistemological implications arise.[5] Color, for example, is a hallucinatory artifact which has no basis in physics (as Helmholtz stressed[6]). An interesting second artifact is the now of subjective experience which has the character of an unphysical "more than one-dimensional time"[7] (cf. also Ref. 8).

Thus the problem of the "interface" between brain and world arises (cf. Ref. 9). To understand the nature of this problem, it is necessary to adopt two vantage points simultaneously — that of a hypothesized external (super)observer, and that of the brain-bound window valid for the internal observer. To the internal observer, the danger of an infinite regress lurks because the presupposed totally external (exo) perspective is never accessible to him. Only the first step of the iteration — dubbed the "Copernican turnabout" by Kant[10] — can be performed safely. The "auto" (the "self" of the observer) and the "world" are generated simultaneously.[11]

In the following, this "double thinking" will be extended to the micro realm.

15.2   Space and Time

Space and time are brain-bound, as Boscovich[12] asserted (see Appendix to Chap. 10 for a more modern translation). A simple illustration is provided by the speed of the dynamic processes which go on in the brain. They transform time. To a night-cold lizard who looks for a sunny stone wall, many events appear too fast to be graspable. The example shows that "the impressions generated" (Boscovich's term for "interface") can be the same in two objectively differing situations: cold brain / slow environment and warm brain / fast environment. This principle was first formulated on the basis of another example (presence or absence of slow corotation of observer and world[12]) by Boscovich, who had adopted it from Copernicus.

While all space and time are constructed by the brain, Boscovich's principle of the difference goes still further. It claims that "joint changes" in the brain and the environment can cancel out. The perceived virtual reality depends on both the properties of the external world and the properties of the observing system, on an equal footing. Two objectively different spatiotemporal situations can therefore become identical as far as the interface is concerned. To a night-cold lizard, a slow-moving environment has the same speed as a fast-moving environment has to a sun-warm lizard.

The power of this "equivalence principle" is still uncharted. To get an idea of its potential scope, one might picture a completely unfamiliar space–time (a perfectly scrambled one) and realize that, depending on the way the cut between a subsystem and the rest of that universe is chosen, a completely conventional space–time can come out in the interface. Conversely, one could start out from a thoroughly familiar space–time and generate a complete mess ("tace and spime") in the interface — as we shall see.

15.3 Constructivism and Holism

Constructivism is a form of holism. It is an attempt to reconstruct the whole starting from a cut. Hereby the methodological procedure adopted often goes the other way around, though. One starts out "heuristically" positing a certain absolute reality. Then one derives properties of the internally existing interface and posits them as possible predictions about reality. An example is Uexküll's distinction between "objective environment" (*Umwelt*, literally "around-world") and "effective environment" (*Wirkwelt*, "affecting world"). One of the suggestive insights derived in this fashion was, for example: "The jelly-fish only hears the toll of his own bell"[14] (cf. Ref. 5).

The "radical constructivism" in biology, in the sense of von Foerster,[15] Watzlawick[15a] and von Glasersfeld,[15b] and the "holistic thinking" in physics, in the sense of Bohm[16] and Finkelstein,[17] are closely related. The former paradigm is macroscopic and classical in nature, looking for interface-like (counterintuitive) implications in the perceived structure of the world.[17a] The latter paradigm is microscopic and nonclassical in nature, looking for interface-like (counterintuitive) implications in the physical structure of the world. The hope in the second (physical) case is to derive features of the empirical quantum reality from the presupposed ("hidden variables") reality (which involves either nonstandard force fields[18] or nonstandard logics[17]). The methodological analogy between biological constructivism and physical holism permits one to use the same term for the two approaches: "constructivism." Classical biological constructivism then becomes "macroconstructivism," and physical holism becomes "microconstructivism."

A third example in the micro class, besides Bohm's and Finkelstein's non-classical proposals, would be the classical approach proposed by Popper.[19] In the following, a fourth example will be considered which is related to Popper's but has the asset of being immune to a famous counterargument which is believed to disqualify any classical holism.[20]

15.4 Nonlocality in Microconstructivism

One of the aims of microconstructivism is the rederivation of some (or all) features of the quantum reality as interface-generated properties. For example,

Bohm's deterministic hidden variables approach to quantum mechanics (the so-called quantum potential[18]) successfully reproduces every observable property of quantum mechanics. Nevertheless it should be mentioned immediately that the quantum potential lacks an explicit internal observer and an interface (a fact which has far-reaching implications, as we will see soon). In 1964, Bell[20] discovered that Bohm's hidden variables theory[18] fails to explain an important feature of quantum mechanics: its "nonlocality." Nonlocality consists in an apparent superluminal influence, exerted by one measurement on the outcome of a distant other measurement if the latter is performed on a "twin" particle.[20] The reason for the failure of Bohm's theory lies — superficially speaking — in its closeness to quantum mechanics (so that the property of nonlocality is built in from the beginning, just as it is in quantum mechanics). However, the deeper reason, Bell realized,[20] is that *any* deterministic hidden variables theory which like Bohm's successfully reproduces the quantum nonlocality must be nonlocal itself from the beginning (just like Bohm's).

This famous theorem[20] at first sight spells the end of any microconstructivist approach to the quantum reality provided the latter is based on classical-local hidden variables — like the "Popperian" one considered here. However, a subtle twist prevents Bell's theorem from being applicable to this type of holism. The already mentioned fact that (unlike Bohm's later holism[16] which introduces a cut) Bohm's original hidden variables theory[18] lacks an explicitly specified internal observer and interface suddenly acquires unexpected significance. For this fact has entered Bell's proof as an explicit assumption. Bell's theorem applies only to such hidden variables theories in which the underlying hidden reality (Bell's parameter lambda) produces the nonlocal measurement results "directly." This assumption ("separable predetermination"[20]) is not the rule but rather the exception in microconstructivist theories.

In the general ("Boscovichian") case, the interface reality is not essentially identical with the underlying objective reality but depends on the properties of the observer. Bell's theorem then ceases to be applicable. Hence a classical (both deterministic and local) hidden reality *can* in principle reproduce the nonlocal features of quantum mechanics — in the interface.[21] The price to be paid is the same as with any other constructivist explanation: the obtained reality is valid, not objectively but only as an interface-generated "illusion."

15.5 Kinetic Constructivism

A first explicit local example of the microconstructivist approach is "kinetic constructivism." The postulated hidden level of reality here is identical with kinetic theory — Boltzmann's classical-mechanical universe of elastic billiard balls. (Hereby charges and electromagnetism can be formally added, as Wheeler and Feynman[22] showed.) Chaos theory has made this mathematical universe attractive again.[23,24]

There exist three little-known results in this class of systems which deserve to be mentioned here because of their relevance to the interface problem.

The *first* unfamiliar implication of kinetic theory is the in-principle-well-known fact that any classical gas at equilibrium occupies a finite phase space volume. At equilibrium, phase space volume becomes maximal. Phase space volume is defined as the "extension" in all directions of phase space (both as far as position and as far as momentum is concerned) shown by all particles. Since the product of a position and a momentum (velocity times mass) is an action according to Leibniz, the phase space volume has the dimension of an action raised to the $3N$th power (if there are N particles moving in 3-space).[25] This old result of Gibbs implies that a "unit action" exists in any closed classical system. This action is equal to the $3N$th root of phase space volume. This formal result is for some reason not very well known. It would be of special significance if the so-obtained action were representative for an average particle and if — in addition — its magnitude were not unrealistically large as one expects it to be at first sight. Both desirable features turn out to be realistic.

The described "classical unit action" is indeed surprisingly small if the gas or liquid in question contains mathematically equal particles (of only a few types). This, however, is the general case. Note that particles can be considered as soliton-like solutions to a classical field[26] (solitons are in the simplest case isolated traveling waves that are indestructible). The assumption of mathematical equality of physical particles is therefore plausible (and is in fact undoubted). The consequence is that phase space volume is reduced by a giant factor. The reduction factor is equal to $n_1! \cdot n_2! \cdots$ (if n_1 is the number of equal particles of the first type and n_2 that of equal particles of the second type, and so forth[25]).

For example, if more than half the particles belong to the lightest class endowed with the mass of the electron, and if the density and the temperature of biology's semiliquid materials (brains) is assumed, one obtains for the unit action specific to the "electronic" subdimensions of phase space the value

$$h^* \approx \frac{h}{4\pi}, \tag{1}$$

up to a factor of less than 2, where h is Planck's constant.[27] This result (implicit in the formula for P of Chap. 2) is independent of any openness or closedness (far from or close to equilibrium) conditions.[28]

It goes without saying that the significance of the "almost-coincidence" of Eq. (1) (where the constant h^* is not universal but depends on density and temperature) is currently unknown. This situation might change if it turned out that the classical action h^* can be "tuned through" experimentally such as to give rise to new quantum effects; cf. Ref. 28a.

A *second* mathematical implication is the existence of a finite time interval, T, the mean passage time of the phase space cell. It is of the order of

$$T \approx \frac{h^*}{E}, \tag{2}$$

where E is the unit thermal noise energy (1/2 times Boltzmann's constant times temperature) valid in the system in question.[29]

The *third* mathematical implication is a corollary to the second. Every unit time interval T', where

$$T' \approx T, \tag{3}$$

an "effective time reversal" occurs. More specifically, the forward and the backward motions in the chaotic configuration space (think of a point-shaped billiard ball moving on a frictionless table as the simplest illustration) have equal densities locally. This property goes hand in hand with the fact that trajectorial segments which cross the phase space cell belong to two equivalence classes which are mutually identical under time reversal. As a consequence, "most" projections have inverted their temporal orientation after T'; see Ref. 29a.

15.6   Consequences for the Interface

The preceding three results, Eqs. (1)–(3), look as if they might have the capacity to generate an "analog" to quantum mechanics in a classical-kinetic universe. Three specific conjectures would need to be fulfilled if one really wanted to demonstrate such a result to be valid for the interface between a far-from-equilibrium observer and the rest of his or her kinetic universe:

(I) The action h^* enters as a "perturbation" any micro measurement as represented in the interface. ("Quantum noise.")

(II) Two identical perturbations apply in the case of two objectively correlated particles. ("Quantum nonlocality.")

(III) Stable, macroscopically represented measurement results form in the interface. ("Quantum eigenstates.")

The odds for all three conjectures to be demonstrable appear to be negligible at first sight. Unexpectedly, there is a way it appears if the main implication of the preceding section, Eq. (3), is taken as the point of departure.

Equation (3) implies a radical transformation of the world in the interface. The reason lies in Boscovich's equivalence principle, which includes the following special case: a time reversal in the observer is equivalent to a time reversal in the external world without a time reversal in the observer (see Chap. 10). As a consequence, Eq. (3) implies the existence of a "micro vacillation of causality" in the external world, as far as that world is represented in the interface.

What properties does a causally vacillating world possess?

15.7   A Strange–Strange World

The alternating time reversals in the external world, every second time interval T', do not affect all events in the external world to an equal extent. Only those events which directly reach the observer (via a "measuring chain") are involved. For those events, on the other hand, the measuring chain reverses its causal orientation in time over its whole length. This is because every causal effect which contributes to the behavior of the final element of the measuring chain, as it impinges on the observer, is also linked — during the next time slice T' — with the opposite behavior of the same final element.[29]

It follows — if at least two successive time slices are considered jointly — that all observable objects are connected with the observer in a kind of "two-way causality." Hence a "reversible" (more specifically, action-conserving) coupling is implied.[29] This coupling bridges the whole (both spatial and temporal) length of the measuring chain. Hence the observer "perturbs" the whole observable world across all causal chains that connect any object to him.[29] Since the different measuring chains have temporal extensions that vary at random, the momentarily valid "world on the interface" is characterized by a "fractal-like distortion" of the entire world including its past.

Conjecture (I) of the preceding section — action perturbation — is thus correct.

The same holds good for conjecture (II) — correlated perturbations — since it forms a special case of the former. For it refers to the observation (with the aid of two measuring chains) of a single object — the originally composite particle as it splits into two subobjects that both allow one to measure its state at the moment of splitting (cf. Chap. 8).

There remains conjecture (III) — stable measurement results. The distortions of the whole world, generated by the causality vacillation, are by definition highly transitory ("fleeting"). Conjecture (III), however, explicitly demands that they be nonfleeting.

Thus an "impasse" appears to have been reached. This impasse comes in the wake of an earlier one, implicit in Eq. (1) but not yet stressed, that h^* depends on the body temperature of the observer and therefore is not a universal constant like h. Hence a seemingly "hopeless" situation has occurred. In such a case it can be a good strategy to continue with the main thread of the story as if nothing had happened — in the hope that it will yield a clue. We accordingly return to the strange world of a causally vacillating observer and ask ourselves the question: What are the most incisive implications of such a world? They are:

(i) The behavior of micro objects, as far as they are represented in the interface, is illusory (exo-objectively nonexistent).

(ii) The illusory character cannot be unmasked directly by the internal observer.

(iii) Even the transitory character of the momentary illusions is not accessible to the observer.

Feature (iii), if correct, solves the "impasse" — by postulating a "masking" effect. We are therefore allowed to continue if we manage to derive the three principles (i)–(iii) as results.

15.8 Worlds-Hopping

The constructivist paradigm "deconstructs" reality by explaining it as an illusion (a mere interface property). Principle (i) of the preceding Section only restates this fact for *micro*constructivism. It therefore is correct.

Principle (ii) asserts that "double checking" on the content of the interface is more difficult in microconstructivism than in the traditional macroconstructivism. The rationale for this claim lies in the fact that, unlike macro states, micro states of the observer cannot be preserved at will. From this fact it follows that two strategies are blocked now which were open in macroconstructivism. When it comes to correcting for an optical illusion on the macro level, or for a phantom limb sensation, to mention only two examples, it is always possible to resort to conventional macroscopic reasoning and to rely on the judgment of other observers, respectively. Both these strategies are blocked in the micro case. It follows that the deconstruction of properties of the micro interface is more difficult than that of properties of the macro interface. Principle (ii) is an expression of this fact. It therefore is correct.

Principle (iii), finally, can be derived most readily by showing that it forms a special case of principle (ii). In view of its central importance, however, one would rather like to understand it directly — this even more so because principle (iii), in addition to solving the problem of stable eigen states (the above "impasse") as mentioned, also solves the problem of the observer temperature dependence of h^* (the above-mentioned "earlier impasse"). For it goes without saying that, if changes in the representation of individual events in the interface are inaccessible, changes in the mean values of the same representations must be inaccessible, too.

Principle (iii) thus deserves to be looked at in its own right. It reintroduces into physics a property which had been absent since the time of Heraclitus: invisible change.[30,30a] Therefore what remains to be shown is: *Why* is the change invisible?

It is invisible because the interface contains a whole world (complete with recorded past and expectable future). Whenever a whole world changes, however, nothing changes. For a *whole world* by definition contains no trace of another whole world. Two examples may make this clear.

The first example is the universal ancient idea of "transmigration of the soul" (Greek, Eastern, American). It always comes combined with the characteristic assertion that every segment in the soul's history lacks all memory of the previous incarnation. This claim at first sight only serves to make the idea unfalsifiable. However, at second glance one realizes that the assertion is a logical implication of the whole-world principle.

The second example is a little-known version of quantum mechanics that was discovered by Bell.[31] It assumes that all the "relative states" of quantum mechanics described by Everett[32] exist, not simultaneously along a new extraphysical dimension (as Everett's theory is commonly understood) but sequentially — along the time axis (as mentioned in Chap. 13). The "transmigration" thereby involves not just five or ten lives this time but almost infinitely many — since every segment lasts only for a near-infinitesimally short time interval (the fortieth part of a picosecond, say). Paradoxically, there indeed exists no lower bound here: the duration of each segment is a matter of complete irrelevance, provided the individual segments are internally consistent each.

This playful idea of Bell's is re-encountered here — not as an illustration of ancient thinking or as a curiosity in the history of quantum mechanics, but as an implication of microconstructivism. We saw that if principles (i)–(iii) hold good, the three main properties of quantum mechanics (noise, nonlocality, stability) are recovered since they then, by conjectures (I)–(III), govern the interface reality of a kinetic observer. Hence quantum mechanics has been reobtained, it appears. (Only quantitative details — like the derivation of the Schrödinger equation[29] — are still wanting.) The version obtained, however, is Bell's.

15.9   Assignment

Kinetic constructivism combines, as we saw, a "many-world" theory (Bell's) with a "single-world" theory (Boltzmann's). As a consequence of this unification, a new element of physical reality can be defined: assignment conditions.

To describe nature completely, it no longer suffices to specify the "laws" and the "initial conditions" (as Newton did). It in addition becomes necessary to specify the "assignment conditions." Newton and Einstein were well aware of their existence: those initial conditions which refer to a particular piece of matter (one's own brain) or to a particular relativistic frame (containing that piece of matter) are special. However, this fact does not show up in the mathematical description of nature. Due to the fact that only a macroscopic lump of matter seemed to be involved, no surprises were expected.

Microconstructivism changes this complacent view. The world generated in the interface never exists for more than one "now." This "now" is connected to a particular micro state of the observer. The world valid at a given "now" depends on that micro state. It would be a different world if another micro state had been picked. This holds good even when the new micro state possesses exactly the same dynamics as the old one. In other words, there exists a type of micro state on which the world depends but which makes no difference in the universe. The new micro state is determined, not by the dynamics of the universe but by microscopic "assignment decisions."

They declare: "This particular electron," characterized by its unique spatial cell,[28a,34] is to belong to the "observer" and "that one" to the "rest of the universe." The interface valid for the observer depends critically on this micro assignment.

The general importance of assignment was apparently last appreciated in science at the time of the outdated "syncytial theory" of the nervous system. This theory claimed that a single cell, with many nuclei but bounded by a single membrane (a so-called "syncytium"), makes up the conscious subportion of matter in the skull.[35] This theory was disproved by the invention of the electron microscope.[35a] Nevertheless it was a grandiose macroscopic theory.

Now a "microscopic" assignment appears on stage. Micro assignment would be a new, most intimate activity of nature.[36] Its existence in the real world is contingent on the validation of microconstructivism (cf. the next section). However, the problem of "which interface" to choose arises also in computer simulations already.

If one has a "billiard-ball universe" implemented in one's computer — as this is possible today[36a] — a large number of distorted interface-bound projections become available in principle. (For the time being, the rules for calculating the interface for a more-than-single-particle observer are not yet explicitly available.) All possible micro interfaces then wait in line to be put on display on the screen. This selection has to be made, after the dynamics of the pertinent universe and the macroscopic identity of the internal "ID unit" of interest as well as the internal "moment" of interest have been fixed.[37] There appears to be no way around this microscopic task if a realistic version of what holds good for the internal ID unit at a given moment in time is to be brought to "life" in the case of a microscopically simulated universe. Thus a "many-interfaces theory" becomes as unavoidable here as the "many-minds theory" of David Z. Albert[37a] (a version of Everett's theory) does in real physics. The notion of the interface thus indeed opens up a whole new degree of freedom — that of assignment — when it comes to dealing with a microscopically specified universe.

15.10   Testable Predictions

The computer renders microconstructivism operational, as we saw. It permits one to look at gadgets (explicit observers: cf. Chap. 8) that are part of a complete universe and to watch their predicament when confronted with the task of understanding their own world. Although the day when one can actually talk to them lies still in the distant future, some of the relevant questions can already be formulated today. The same questions can be applied directly to our own world — without having been tested in the computer before.

Three different types of experiment can in principle be done by the inhabitants of a classical-kinetic universe, once they have become aware of the interface problem. The first type's aim is to expose the "existence" of the interface. The second's aim is to "leave" the interface in an "indirect" manner — by manipulating the hidden level of objective reality in a "blindsight" fashion. (The name derives from the analogy with macro experiments in which human beings who are blind because of a defect in the visual cortex nevertheless successfully catch a large ball thrown to them, for example.[37b]) The third type's aim is to leave the interface in a "direct" manner — by gaining access to the assignment conditions themselves. These three classes of experiment are performable in the artificial universe in principle. Their "analogs" can be performed in the real world.

An example in the *first* class is the "relativistic Bell experiment," which can be done today with the aid of a satellite.[38] As mentioned in Chap. 10, it generalizes a proposal made by Einstein, Podolsky and Rosen,[39] replacing the nonrelativistic — mere "spacelike" — separation between two correlated particles proposed by these authors by a more efficient relativistic — "causal" — separation. The outcome of the experiment, persistence of the Bell correlations, is not in doubt.[40] It will prove that a single quantum world is not sufficient because a different one is needed for the other frame. This amounts to an unexpected confirmation of Einstein's claim that quantum mechanics is "incomplete."[39] More important, however, the experiment provides strong evidence in favor of the existence of an observer-specific interface.

An experiment in the *second* class is almost performable to date. It is a test of covariance between rotational frames.[41] A slow Copernican corotation should be undetectable from the inside of a sealed lab in a two-dimensional universe in accord with Boscovich's 1755 claim.[12] There exists a well-known quantum experiment in three dimensions — absolute superfluid nonrotation[42] — which will permit an analogous decision when repeated at greater accuracy. Absolute superfluid nonrotation is believed to be "genuinely" absolute.[43] The "positive" outcome predicted by two-dimensional microconstructivism (corotation with the observer) would imply the existence of a new nonlocality (counterfactual change of records at a distance). The "negative" outcome predicted by the Copenhagen version of quantum mechanics (although not by other versions of quantum mechanics, like Everett's[43a]) will falsify two-dimensional microconstructivism.

The *third* class of blindsight experiments performable in an artificial universe is devoid of a worked-out analog for the time being. Nevertheless there exists an empirical approach worth mentioning here because it can be interpreted in an analogous manner. This approach is called "micro psychophysics."[44] The microscopic physical state of a conscious observer's brain is here manipulated through the application of external magnetic fields, for example. Even mere confirmation of the predictions made by quantum mechanics will already be of interest since Fechner's method[45] has apparently not been used before as a method in biological materials science. The drawback lies on the theoretical side. Since charged particles cannot yet be included in kinetic constructivism, specific changes in the interface to watch for under the innocuous experimental conditions of this method cannot be predicted as yet. Effects not yet covered by theory, however, have only a low chance to be discovered experimentally.

15.11 Conclusions

Microconstructivism is a theory of the brain. At the same time, it enables a new theory of physical reality. A classical kinetic universe is assumed, on the outside. Then, on the inside, a whole quantum-like reality is implicit — as an "interface reality."

Postmodern deconstructionism thereby enters physics and brain theory. Simultaneously, the notion of transmigration of the soul acquires scientific acceptance. Is this still physics?

Explanation is a voluntary endeavor — a new type of explanation can always be rejected if one feels safer in the fold of the old paradigm (even if this amounts to renouncing explanation in the eyes of the new paradigm). The question of whether microconstructivism indeed is a paradigm is currently premature to ask. The greatest weaknesses of microconstructivism still hold its greatest promise. The two weakest points are, (1) the observer-centeredness of the micro interface and, (2) the idea of micro assignment that goes with it. The potential payoff lies in the implied prediction that simultaneity has a fractal-like structure. The whole world — along with past and future — would be transformed in a nonlinear fashion at every moment. The fractal "mist" would rise and set, in rapid succession, without this fact being recognizable. Nowness would become an integral part of physics. Eventually, even a manipulation of the now becomes envisionable — through reaching "behind appearances" in specially designed experiments.[46] It is therefore perhaps fortunate that the observer-centeredness of the interface is so strong a prediction that its experimental falsification will not be difficult.

A "biologist's physics" has been presented. The brain may play an active role in constructing the world, not only by virtue of its macroscopic properties (auditory and visual cortex, etc.) and its macro dynamics (including chaotic itinerancy[47]), but also through its micro dynamics. The latter may influence the spatiotemporal structure of the world, not only "subjectively" (in the corrigible fashion of macroconstructivism), but also "objectively."

In this way a new notion of objectivity is introduced into physics. It is the "fake objectivity" which Kant discovered in philosophy. All we can know is that the things "in themselves" are different from the way they appear to us within the categories of space and time and causality, Kant taught.[10] While this is as far as philosophy can go, the job of physics is easier because it can construct its own things in themselves. The latter are not the real ones but play their roles "as far as the system is concerned" (namely, the system of physics).[48] Physics therefore has an easier life than philosophy: it can posit its own absolute world heuristically. The infinite regress is thereby considerably defused. This idea leads first to constructivism in the traditional (macro) sense and then to microconstructivism.

Microconstructivism suffers from two general drawbacks. The first is a direct consequence of the fact that from a philosophical point of view, it is only a first approximation. Therefore the first round of physical experiments has very little chance to succeed. The corresponding good news is that no matter how miserably the approach will fare on empirical grounds, nothing will be lost on the part of philosophy.

The second general drawback is — of all things — quantum mechanics. While a version of quantum mechanics can apparently be derived as a micro interface phenomenon as we saw, quantum mechanics is not just sitting there waiting to be explained: it actively fights back. The decisive argument is due to von Neumann.[49] He already glimpsed the microconstructivist danger (so to speak) when he wrote that the limited "state of information of the observer regarding his own state" might be responsible for the quantum indeterminacy. In the next few sentences, he was able to show that the formalism of quantum mechanics rules out the possibility that knowledge (or not) of the quantum state of the observer makes a difference for the outcome of any measurement.

This assertion of von Neumann's contradicts Everett's later theory.[43a] The negative outcome of the rotational covariance experiment in 3D[49b] is in accord with 3-D microconstructivism (Chap. 14 and Ref. 50). Next, a variant of the 3-D experiment that is genuinely 2-D has to be sought. The genuinely 2-D quantum Hall effect[51,52] can apparently be used to obtain an experimental decision.

To conclude, the deconstruction of the world implicit in biological constructivism can be extended to the micro realm. The simplest illustration is provided by classical-kinetic theory. The notion of the "observer–rest-of-the-universe interface" provides a link between physical holism in the sense of Boscovich and the modern technology of virtual reality.[53] However, there is an even older root: the ancient Yin Yang picture.

Acknowledgments

I thank Kuni Kaneko, Ichiro Tsuda, Kazuhiro Matsuo, Michael Conrad, Alistair Mees, Norman Packard, Walter Freeman, Okan Gurel, Oktay Sinanoglu, Paul Mezey, Bill Smith, Bill Langford, Harry Swinney, Doyne Farmer, Jim Crutchfield, George Kampis, Peter Erdi, Bob Rosen, Georg Franck, Hanns Ruder and Peter Weibel for discussions. I also thank Nils Röller for a thorough reworking of the manuscript. Work supported in part by the DFG. For J. O. R.

References

1. E. C. Zeeman, "Topology of the brain," in *Mathematics and Computer Science in Biology and Medicine*, (Proceedings of Conference, Medical Research Council), pp. 277–292 (Her Majesty's Stationery Office, London, 1965).

1a. O. Gurel, "Topological dynamics in neurobiology," *Int. J. Neuroscience* 6, 165–179 (1973); "Hierarchical oscillations," in *Proc. 13th Int. Conf. Chronobiology 1977* (F. Halberg, ed.), pp. 325–332 (Milano: IT Press, 1981).

2. I. Tsuda, "A hermeneutic process of the brain," *Prog. Theor. Phys. (Suppl.)* 79, 241–259 (1984).

3. W. Freeman, "Simulation of a chaotic EEG pattern with a dynamic model of the olfactory system", *Biol. Cybernetics* 56, 139–150 (1987).

4. O. E. Rossler, "Adequate locomotion strategies for an abstract organism in an abstract environment — a relational approach to brain function," in *Physics and Mathematics of the Nervous System* (M. Conrad, W. Güttinger and M. DalCin, eds.) (Berlin: Springer-Verlag), *Lecture Notes in Biomathematics* 4, 342–369 (1974); "Deductive biology — some cautious steps," *Bull. Math. Biol.* 40, 45–58 (1978); "Artificial cognition-plus-motivation and hippocampus," in *Neurobiology of the Hippocampus* (W. Seifert, ed.), pp. 573–588 (New York: Academic, 1983).

5. K. Lorenz, *Behind the Mirror* (London: Methuen, 1977).

6. H. L. F. von Helmholtz, *Treatise on Physiological Optics* (J. Southal, transl.) (New York: Dover, 1925), first edn., 1868; "The facts in perception (1878)," in *Helmholtz on Perception* (R. M. Warren and E. P. Warren, eds.), pp. 205–246 (New York: Wiley, 1968).

7. H. A. C. Dobbs, "The dimensions of the sensible present," in *The Study of Time* (J. T. Fraser and F. C. Haber. eds.), pp. 274–292 (Berlin: Springer-Verlag, 1972); R. Salamander, *Zeitliche Mehrdimensionalität als Grundbedingung des Sinnverstehens* ("Temporal Higher-Dimensionality as a Basic Condition for Understanding Meaning") (Berne: Peter Lang, 1982); G. Franck, "Virtual time — can subjective time be programmed?", in *Ars electronica 1990: Virtual Worlds* (G. Rattinger, M. Russel, C. Schöpf and P. Weibel, eds.), pp. 57–81 (Linz: Veritas-Verlag, 1990); "Internal time and temporality," in *Inside Versus Outside — Endo- and Exo-Concepts of Observation and Knowledge in Physics, Philosophy and Cognitive Science* (H. Atmanspacher and G. J. Dalenoort, eds.) (Berlin: Springer-Verlag, 1994), pp. 63–83; S. Vrobel, *Fractal Time* (Houston: International Association of Interdisciplinary Research, 1997).

8. O. E. Rössler, "An artificial cognitive-map system," *BioSystems* 13, 203–209 (1981).

9. H. von Foerster, "On constructing a reality," in *Environmental Design Research*, Vol. 2 (F. E. Preiser, ed.), pp. 35–46 (Stroudberg: Dowden, Hutchinson and Ross, 1973).

10. I. Kant, *Critique of Pure Reason* (N. Kemp Smith, transl.) (New York: St. Martin's Press, 1929), preface to 2nd 1787 edn.

11. H. R. Maturana and F. J. Varela, *Autopoesis and Cognition — The Realization of the Living* (Dordrecht: Reidel, 1980). 1st Spanish edn.. *De maquinas y serez vivos* (Santiago: Editorials Universitaria, 1972); *El árbol del conocimiento* ("The Tree of Knowledge") (Santiago, 1984).

12. R. J. Boscovich, "On space and time as they are recognized by us," in *R. J. Boscovich — A Theory of Natural Philosophy* (J. M. Child, ed.), pp. 203–205 (Cambridge, Mass.: MIT Press, 1976) (a 1922 translation of the 1755 Latin original).

13. O. E. Rössler, "Boscovich covariance," in *Beyond Belief: Randomness, Prediction and Explanation in Science* (J. L. Casti and A. Karlqvist, eds.), pp. 69–87 (Boca Raton, Fla.: CRC Press, 1991). Cf. Chap. 10 of the present book.

14. J. von Uexküll, *Theoretische Biologie* ("Theoretical Biology") (Berlin: Julius Springer, 1920), reprint of 2nd 1928 edn. (Frankfurt: Suhrkamp, 1973).

15. H. von Foerster, *Observing-Systems* (Seaside, Cal.: Intersystems, 1981).

15a. P. Watzlawick, *How Real Is Reality?* (in German) (Munich: Piper, 1976).

15b. E. von Glasersfeld and J. Richards, "The control of perception and the construction of reality," *Dialectica* **33**, 37–58 (1979).

16. D. Bohm, *Wholeness and the Implicate Order* (London: Routledge and Kegan Paul, 1980).

17. D. Finkelstein, "Holistic methods in quantum logic," in *Quantum Theory and the Structures of Time and Space*, Vol. 3 (L. Castell, M. Drieschner and C.F. von Weizsäcker, eds.), pp. 37–60 (Munich: Carl Hanser, 1979).

17a. D. D. Hoffman, "The interpretation of visual illusions," *Scientific American* **249**, 6, 154–162 (1983); cf. also the formulation of a macroscopic "observer theory": B. M. Bennett, D. D. Hoffman and C. Prakash, *Observer Mechanics — A Formal Theory of Perception* (San Diego: Academic, 1989).

18. D. Bohm, "A suggested interpretation of the quantum theory in terms of hidden variables I, II," *Phys. Rev.* **85**, 166–179, 180–193 (1952).

19. K. R. Popper, "Indeterminism in classical physics and quantum physics I," *Brit. J. Phil. Sci.* **1**, 117–133 (1951), p. 129; "Autobiography," in *The Philosophy of Karl Popper* (P. A. Schilpp, ed.), Vol. 1, pp. 3–181 (LaSalle: Open Court, 1974), pp. 102, 103.

20. J. S. Bell, "On the Einstein–Podolsky–Rosen paradox," *Physics* **1**, 195–200 (1964); reprinted in: J. S. Bell, *Speakable and Unspeakable in Quantum Mechanics*, pp. 14–21 (Cambridge: Cambridge University Press, 1987).

21. O. E. Rössler, "Explicit observers," in *Optimal Structures in Heterogeneous Reaction Systems* (P. J. Plath, ed.), pp. 123–138 (Berlin: Springer-Verlag, 1989). Cf. Chap. 8 of the present book.

22. J. A. Wheeler and R. P. Feynman, "Interaction with the absorber as the mechanism of radiation," *Rev. Mod. Phys.* **17**, 157–162 (1945).

23. Ya. G. Sinai, "Dynamical systems with elastic reflections," *Russ. Math. Surveys* **25**, 137–190 (1970).

24. O. E. Rössler and M. Hoffmann, "Quasiperiodization in classical hyperchaos," *J. Comp. Chem.* **8**, 510–515 (1987).

25. J. W. Gibbs, *Elementary Principles in Statistical Mechanics* (New Haven: Yale University Press, 1902), Chap. 15.

26. D. Finkelstein, *J. Math. Phys.* **7**, 1218 (1966); T. H. R. Skyrme, *Nucl. Phys.* **31**, 556 (1962).

27. O. E. Rössler, "An estimate of Planck's constant," in *Dynamical Phenomena in Neurochemistry — Theoretical Aspects* (P. Érdi, ed.), pp. 16–18 (Budapest: Publications of the Institute of Theoretical Physics of the Hungarian Academy of Sciences, 1985).

28. O. E. Rössler, "Indistinguishability implies quantization," in *Collected Papers Dedicated to Professor Kazuhisa Tomita on the Occasion of His Retirement from Kyoto University* (S. Takeno, T. Kawasaki and H. Tomita, eds.), pp. 280–288 (Kyoto: Publication Office of *Prog. Theor. Phys.*, 1987).

28a. O. E. Rössler, "Symmetry-induced disappearance of reality — the Leibniz effect," *Leonardo* **22**, 55–59 (1989).

29. O. E. Rössler, "A possible explanation of quantum mechanics" (preprint, 1985), in *Advances in Information Systems Research* (G. E. Lasker, T. Koizumi and J. Pohl, eds.), pp. 581–589 (Windsor, Ont.: International Institute for Advanced Studies in Systems Science and Cybernetics, University of Windsor, 1991).

29a. O. E. Rössler, H. H. Diebner and W. Pabst, "Micro relativity," *Z. Naturforsch.* **52a**, 593–599 (1997).

30. O. E. Rössler, "Into the same rivers we step and step not, we are the same and we are not — on the origin of the now," German translation in: O. E. Rössler, *Endophysik — Die Welt des inneren Beobachters* ("Endophysics — The World of the Internal Observer") (Berlin: Merve-Verlag, 1992), pp. 157–179. Cf. Chap. 12 of the present book.

30a. O. E. Rössler, *Das Flammmenschwert* ("The Flaming Sword") (Berne: Benteli-Verlag, 1996).

31. J. S. Bell, "Quantum mechanics for cosmologists," in *Quantum Gravity 2* (C. Isham, R. Penrose and D. Sciama, eds.), pp. 611–637 (Oxford: Clarendon, 1981); reprinted in: J. S. Bell, *Speakable and Unspeakable in Quantum Mechanics*, pp. 117–138 (Cambridge: Cambridge University Press, 1987).

32. H. Everett, III, "'Relative-state formulation' of quantum mechanics," *Rev. Mod. Phys.* **29**, 454–462 (1957).

33. O. E. Rössler, "Endophysics," in *Real Brains, Artificial Minds* (J. L. Casti and A. Karlqvist, eds.), pp. 25–46 (New York: North-Holland, 1987). Cf. also Chap. 6 of this book.

34. O. E. Rössler, "A chaotic 1-D gas — some implications," *Lecture Notes in Physics* **278**, 9–11 (1987).

35. G. Ricker, *Relationspathologie — Pathologie als Naturwissenschaft* ("Relational Pathology — Pathology as a Natural Science") (Berlin: Springer-Verlag, 1924).

35a. E. Letterer, *Allgemeine Pathologie, Grundlagen und Probleme, ein Lehrbuch* ("General Pathology, Foundations and Problems — A Textbook") (Stuttgart: Georg Thieme Verlag, 1959).

36. O. E. Rössler and J. O. Rössler, "The endo approach," *Applied Math. & Computation* **56**, 281-287 (1993).

36a. H. H. Diebner and O. E. Rössler, "Deterministic continuous molecular dynamics simulation of a chemical oscillator," *Z. Naturforsch.* **50a**, 1139-1140 (1995).

37. D. F. Galouye, *Simulacron Three*, 1964. German translation: D. F. Galouye, *Welt am Draht* ("A Puppeteer's World") (Munich: Heyne-Verlag, 1965).

37a. J. Horgan, "Quantum philosophy," *Scientific American* **267** (July), 72-80 (1992).

37b. L. Weiskrantz *et al.*, "Visual capacity in the hemianopic field following a restricted cortical ablation," *Brain* **97**, 709-728 (1974).

38. O. E. Rössler, "Einstein completion of quantum mechanics made falsifiable," in *Entropy, Complexity and the Physics of Information* (W. H. Zurek, ed.), pp. 367-373 (Redwood City: Addison-Wesley, 1990).

39. A. Einstein, B. Podolsky and N. Rosen, "Can quantum mechanical description of physical reality be considered complete?", *Phys. Rev.* **47**, 777-780 (1935).

40. O. E. Rössler, "Bell's symmetry" (preprint, 1990).

41. O. E. Rössler, R. Rössler and P. Weibel, "'Absolute' superfluid nonrotation: is it observer-frame-specific?" (preprint, 1991); *Die Welt als Schnittstelle* ("The world as interface"), in *Vom Chaos zur Endphysik* (F. Rötzer, ed.), pp. 369-381 (Munich: Klaus Boer Verlag, 1994).

42. G. B. Hess und W. M. Fairbank, "Rotation of superfluid helium," *Phys. Rev. Lett.* **19**, 216-219 (1967).

43. R. E. Packard and S. Vitale, "Principles of superfluid helium gyroscopes" (Berkeley preprint, 1992).

43a. O. E. Rössler, "Relative-state theory: four new aspects," *Chaos, Solitons and Fractals* **7**, 845-856 (1996).

44. O. E. Rössler and R. Rössler, "Is the mind-body interface microscopic?", *Theoretical Medicine* **14**, 153-163 (1993).

45. G. T. Fechner, *Elemente der Psychophysik* ("Elements of Psychophysics") (Leipzig: Breitkopf und Hästel, 1860).

46. O. E. Rössler, *Ist unsere Welt eine virtuelle Realität?* ("Is our world a virtual reality?") in *Cyberspace* (F. Rötzer and P. Weibel, eds.), pp. 256-266 (Munich: Klaus Boer Verlag, 1993); "Endophysics — Descartes taken seriously," in *Inside Versus Outside — Endo- and Exo-Concepts of Observation and Knowledge in Physics, Philosophy and Cognitive Science* (H. Atmanspacher and G. J. Dalenoort, eds.), pp. 153-169 (Berlin: Springer-Verlag, 1994).

47. I. Tsuda, "Chaotic itinerancy as a dynamical basis of hermeneutics in brain and mind," *World Futures* **32**, 167-184 (1991); K. Kaneko, *Physica* **D41**, 137-172 (1990).

48. I. Kant, *Opus postumum, Convolutes X and XI*, in: E. Adickes, *Kants Opus postumum* (Berlin, 1920). Quotation from: *Kant, Ausgewählte Schriften. Die Grundlagen des kritischen Denkens* ("Kant, Selected Writings. The Fundamentals of Critical Thinking") (W. del Negro. ed.), p. 399 (Gutersloh: Bertelsmann, 1958).

49. J. von Neumann, *The Mathematical Foundations of Quantum Mechanics* (Princeton, N. J.: Princeton University Press, 1955), p. 438, p. 233 of 1932 German edn.

49b. O. Avanel, P. Hakonen and E. Varoquaux, "Detection of the rotation of the earth with a superfluid gyrometer," *Phys. Rev. Lett.* **78**, 3602–3605 (1997).

50. O. E. Rössler and P. Weibel, "Exo and Endo Symmetry" (in German), in *Jenseits von Kunst* ["Beyond Art"], p. 225 (Vienna: Passagen Verlag, 1997).

51. K. von Klitzing, G. Dorda and M. Pepper, *Phys. Rev. Lett.* **45**, 494 (1980).

52. S. Kivelson, D. H. Lee and S. C. Zhang, "Electrons in Flatland," *Sci. Am.* **274**, 3, 64–69 (1996).

53. P. Weibel and O. E. Rössler, *Endophysics — Neue Perspektiven der Physik* ["Endophysics — Is Physics a Special Case of Media Theory?", in German], in *Documenta X, Kassel, 100 Days — 100 Guests* (C. David, ed.), Day of August 19, 1997 (http://www.mediaweb-tv.de/dx/0819/gaeste_frame.html) (full-length video).

16

Interference Is Exophysically Absent

Coauthor: Mohamed El Naschie

Abstract

Interference is the hardest-to-explain property of our quantum reality. Negative probabilities ("amplitudes") have no classical analog, for averaging can never yield a zero outcome starting from nonzero components. As soon as an endophysical explanation of interference is attempted, on the other hand, the price to pay is that the phenomenon ceases to exist exo-objectively. For it remains valid only in the interface of the observer.

On the interface, interference indeed arises when many different successive interface worlds are overlaid in an integrating fashion. Since the particle makes a different motion in each interface world — hitting a counter in the one and taking off from it (hitting it in negative time) in the other — the net effect can indeed be destructive: a "canceling-out." Motions that either come or go act like quantum amplitudes.

Interference thereby becomes "dual" to state reduction. Whereas interference fuses many successive interface worlds as we saw, state reduction "fixates" a single one in the interface at the expense of all others. World-splitting (state reduction) and world-combining (interference) occur in the interface on the same footing — depending on whether a macroscopic record (which cannot be superposed) is present or not.

The prediction that interference is exophysically absent can perhaps be put to a test. The experiment is called the "unbiased quantum coin." In the original quantum coin experiment proposed by Schrödinger in 1935 (illustrated by the fate of an imaginary cat), "heads" and "tails" were defined by two equiprobable quantum events. In the modified version proposed here, "heads" is caused by an event made rarer than 50 percent by destructive interference and "tails" is caused by an event made more frequent than 50 percent by constructive interference. A deterministic "unbiasing" has then to be added — like letting every heads click but only every tenth tails click pass through by means of a computer program. The resulting "coin" is then fully acceptable as the source of a bias-free random binary sequence.

The "modified quantum ordeal" obtained can possibly be put to a good use — to improve the world on the exo level. While the coin is perfectly fair endophysically, it is "more than fair" exophysically — with its "heads" events. While endophysically there is no difference to the classic Schrödinger cat experiment, the "bonus" given to the rarer outcome ("heads") becomes a real gain exophysically; for the interference-induced "dip" is exophysically absent but the compensating bonus is not. Hence a lottery based on the new unbiased quantum coin would, while being innocuous endophysically, generate more than its share of millionaires exophysically. Even though the gain would be "shielded" from being perceptible endophysically, some subtle trickle-down effect from this "exo-altruism" (since it is a gift without direct returns) cannot be ruled out. In the positive case, the existence of the exo level would be demonstrable. The proposal is related to David Deutsch's idea of subtly exploiting Everett's theory.

Acknowledgments

We thank Said El-Nashaie, Vladimir Gontar, Vladimir Zakharov, Peter Plath, Gerhard Lischka, Siegfried Zielinski, Nils Röller and René Stettler for stimulation. For J. O. R.

17
Second Causation — Assignment Conditions

Summary

The relations which exist in the world were classified by Newton into "laws" and "initial conditions." The two notions jointly define causality. Einstein's relativity and Everett's relative state theory have added the observer's state as a third determinant of nature. If reality is observer-state-relative, however, "assignment conditions" must exist. They generate a cut — the effective forcing function acting on the observer. Since the effective forcing function depends on the observer's micro dynamics, the observer identity must be attributed in a microscopically exact fashion. Hence one subset of the particles in the "brain" of an observer have to be assigned individually to the observing subsystem, the other particles to the rest of the universe. Since the assignment conditions determine the whole world (as it appears on the interface of the observer) in a sensitive fashion, they have at least the same rank as the initial conditions and the laws. Since the laws and initial conditions give rise to ordinary causation, the term "second causation" appears appropriate. The decisive question is: Is the second causation manipulable, too? The answer may turn out to be yes.

17.1 Introduction

Einstein introduced a new perspective into physics — a "relativity" of the objective facts of nature with respect to the observer's state of macro motion. The fundamental nature of this revolution is still incompletely understood. Is the "frame" as important as the "absolute world"? What is the fundamental difference between frames and initial conditions?

Two decades after Einstein had discovered relativity, Bohr discovered complementarity. Although no one knows exactly what he meant by this word,[1] it seems to refer to the fact that an internal observer of a classical world suffers certain limitations when it comes to understanding the world objectively — so that only certain mutually exclusive partial ("complementary") aspects remain accessible to the observer.

Bohr thereby introduced an interface reality once more: Bohr's interface is analogous to the relativistic frame, but is microscopic rather than macroscopic in its nature. The "micro frame" is defined as the effective forcing function valid between observer and rest of the universe.[2] Since the observer's own micro dynamics enters the definition of the micro frame, those particles whose dynamics defines the observer (deep in the brain) are distinguished from the rest. Therefore the existence of microscopic "assignment conditions" is implicit in Bohr's approach.

The most important single question to ask may be the following. Suppose the assignment conditions are a new third determinant of nature (after the laws and the initial conditions): Are they as immune to change as the laws are, or are they as manipulable as the initial conditions are? And, if the latter were the case: What are the consequences of such a "manipulation of the second kind"? To find out, it is best to proceed in a stepwise manner.

17.2   Manipulability of the Einstein Cut

Position in space — the determinant of ordinary perspective — is arbitrarily manipulable as an initial condition in the macro realm, as is well known. In the same vein, Einstein's frame — the determinant of the new "perspective" of relativity — appears to be freely manipulable, too. It follows that the mass of a stone and the simultaneity of two events can be changed in defiance of what Newton's laws and initial conditions permit.

This added freedom is limited, however, by what Minkowski's "absolute world" of space–time allows in terms of a possible change of its own initial conditions. Therefore, the basic Newtonian dichotomy appears to be preserved by special relativity since only initial conditions can be manipulated after all. However, general relativity goes one step further. Gödel's[3] demonstration of closed timelike paths in a certain curved space-time shows that a still more radical freedom is implicit: time travel. Time travel (if it were indeed realizable) would dwarf all previous technological achievements of humankind. The effects of a mere change of initial conditions would be transcended. For example, immortality (a control of time) would be among the possible implications.

17.3   Is the Bohr Cut Manipulable as Well?

Micro assignment generates a far more nonlinear space–time hypersurface than general relativity does; cf. Chap. 14. Hence the prospect of still more far-reaching manipulations opens itself up.

Unfortunately, there is a drawback which at first sight appears insurmountable — counterfactuality. While it is true that micro assignment can presumably be manipulated much more readily than macro assignment — by turning around in space or applying a magnetic field to one's head or taking a fever pill (cf. Chap. 8), for example — it goes without saying that nothing changes manifestly under any such condition.

It turns out that even if micro assignment conditions existed and could be manipulated, there should be no way to find out about this fact. The reason has to do with the microscopic nature of the interface. Unlike a macro frame, the micro interface cannot simultaneously be left and retained — in memory, for example. However, any change of which there *is* no record, because all records have changed along, *is* no change.

The same argument applies to time travel. If in Gödel's time machine, the return to the place and moment of departure were "microscopically accurate," there would be no change compared to the first time — it would *be* the first time. An added memory capacity at first sight seems to be the only way to find out about such a state of affairs. The celebrated 1993 movie *Groundhog Day* (Harold Ramis; Bill Murray) conversely features the predicaments of a person endowed with such an added memory capacity.

17.4   Semicounterfactuality

The proposal that the micro interface — the micro assignment conditions — may be manipulable seemed to be unfalsifiable. This fact, if correct, would render the whole idea unscientific. Unexpectedly, there exists a loophole. It could be called semicounterfactuality.

If the whole world depends on the micro interface, there can be no record of a change, as we saw. However, it turns out that a record is not the only way to find out about a change. A roundabout way to make an invisible change manifest exists. The following alternative is in principle open: to choose a micro interface which, despite its hermeticity, "gives away" the fact that the whole world depends on it. The fact that it contains no record of a change is then insufficient to secure hermeticity. The only question is whether or not a case in point can be found.

A hint for finding such a case is implicit in the "method of privileged frames" discovered by Einstein. Privileged frames "must not exist," he postulated. This assumption enabled him to see that the so-called contraction hypothesis of Fitzgerald and Lorentz needed to be replaced by a more symmetric alternative. In other words, the seeming privilege had to be "relativized" (that is, made covariant). The constant speed of light in all directions would then no longer be a sign of privilege but rather a feature shared by all frames in a democratic (covariant) fashion. This program worked out marvelously, as is well known.

The method introduced by Einstein can be "iterated." Specifically, a more complicated type of covariance can be pictured in which an added symmetry shields the covariance from being detectable as easily as this was the case in special relativity. Under such a condition, an actually covariant feature (like c, the speed of light) could seem to be "hermetic." That is, the fact that the same feature applies in another frame would be masked, in each frame. All observers would then be faced with an irreducible privilege once more.

However, the envisaged possibility — that some privileges prove more privileged than others — should not detract from the fact that "search for manifest privilege" is the royal route also to the discovery of this more complicated type of covariance, should it indeed exist.

17.5   Rotational Frame Covariance

A potential example of a "privileged frame" in the new sense is a rotational frame. Specifically, the prediction is possible that spin-free quantum mechanics is "rotational-frame-covariant."[4] An observer who rotates very slowly, along with the whole laboratory including all measuring devices and objects to be observed, should be unable to detect the rotation by means of measuring a corotating 2-D quantum object.

This hypothesis amounts to the prediction of a paradoxical privilege in the above sense. On the one hand, the prediction follows in a straightforward manner from the assumed existence of a classical micro interface. The objective physical world — valid on the interface between the microscopically specified observer and the rest of the universe — is not "absolutely objective" but only "observer-relatively objective," as we saw. Therefore the interface between the observer and the rest (the difference dynamics) is in a 2-D universe bound to be unchanged to first order under the assumed condition of an exo-objective slow corotation.

The new covariance would not be completely hermetic even if all records changed along. This is because the frame-specific privilege introduces a manifest incongruency into physics: a classical gyroscope and a two-dimensional quantum gyrometer would yield mutually incompatible results. It is this fact which makes the proposal semicounterfactual and hence falsifiable.

17.6 Copernicus 1, 2, 3

Copernicus made the first claim of the above type — following in the foot-steps of Nicole Oresme (Nils Röller, personal communication, 1996). The rotation of the earth should cancel out from all purely earthbound observations (as Boscovich[4] later put the idea).

This claim is false on the macro level, as is well known. Foucault's pendulum of 1851 works in hermetically closed rooms, revealing the earth's rotation through maintaining its own plane of oscillation. Similarly, Sagnac's interferometer can be used as a gyroscope with an almost comparable accuracy.

Nevertheless Copernicus' claim can also be checked on the micro level. This is the idea of a "quantum Foucault pendulum."[8] The latter is expected to become available soon and to also become an exceptionally sensitive gyrometer (enabling a better prediction of earthquakes and the discovery of oil layers in the underground, for example). At first sight one would expect that if quantum mechanics is interface-bound, the "absolute zero" of a quantum gyrometer would be invariant to first order in the rotational frame of the laboratory, as we saw. This is because a classical interface — as glimpsed by Copernicus — seems to necessarily possess this property.

It follows at first sight that the "absolute zero" of superfluid nonrotation should be invariant under a condition of very slow corotation with the laboratory compared to the completely rotationless case. Accordingly, in an earthbound laboratory, the measured "absolute zero" should coincide with the rotation rate of the earth on which the observers are at rest, rather than with the absolute zero of the Machian center of the universe.

The prediction made by quantum mechanics, in contrast, is "absolute nonrotation" of the superfluid in its ground state.[5] This prediction has recently been confirmed.[6] The "second Copernican prediction" thus is false, too. How about the third? The third is like the second prediction, but confined to two-dimensional quantum effects. Unlike the second, which is *not* a prediction made by endophysics in 3-D (since the spin model of Chap. 14 leads to a correction which coincides with the prediction made by quantum mechanics), the third prediction appears to be "serious." A corresponding 2-D experiment is in preparation, as mentioned in Chap. 15. The idea that assignment conditions exist and are manipulable thus appears to be falsifiable in principle.

17.7 Conclusions

The concept of "assignment" is very old. One's life and bodily and mental properties and identity are not given into one's own hands.[7] The same applies to the physical world as it is valid at this moment. The current position of the now has "no reason" to it as far as one can tell. Harold Brodkey's (the poet's) sensitivity to the "speed at which the moments pass by" attests to the same predicament.

This "second causation" of assignment is not often considered in science. This is perhaps not surprising in view of the fact that no-one can do anything about it. Moreover, the first causation of events appears to be undisturbed by the second causation. Nonetheless, the second causation has a long history in human thought. Heraclitus called it "*Polemos*" ("War"). "War is the father of all and their king: the ones he declares gods and the others humans, the ones he turns into slaves and the others into free persons" (Fragment No. 53). In the East, the same entity is sometimes called "Mind": "Oh my Mind, once you caused me to be born as a king and then you caused me to be born as an outcast."[8]

In phenomenological philosophy, the whole world also is nothing but a mental structure, a psychic phenomenon, pure consciousness, a "*cogito*."[9] The only really interesting causality is the psyche-assigning "second causation" again. It took a long roundabout way — a "loop," in the terminology of Flusser[10] — to rearrive at this fact in the context of the standard (intramovie) causation defined by Descartes.

The target of the second causation is the interface. More precisely, it is that microscopic substrate in the brain to which also the experiencing consciousness is attached at a given moment.[11] World and soul thereby become connected in a one-to-one fashion. "Soul causation" becomes almost a synonym of "second causation."

Only four physicists after Descartes appear to have taken the idea of an interface-affecting causation seriously: Boscovich, Einstein, Bohr and Everett. Everett's version of quantum mechanics[12] is particularly illuminating. It is completely "observer-relative" in the sense that every measured quantum state in the world is personalized.[2] The world is made "only for me" in a Kafkaesque manner (to mention only the frightening aspect). The now and the quantum world become a single "joint pick" on a two-dimensional parameter surface (Chap. 8).

The "third Copernican experiment" renders Everett's vision falsifiable. Everett's theory thereby ceases to be a mere "interpretation" of quantum mechanics and becomes a full-fledged alternative theory (for the standard theory would still survive after its own downfall). What are the odds for a positive outcome? Unfortunately, they are virtually nil. No prediction at variance with standard quantum mechanics has ever survived empirical scrutiny, with the relativistic Bell experiment[13] (once definitely performed; Chap. 10) forming the only known exception.

Moreover, a positive outcome would have repercussions which appear hardly acceptable. Objective facts of nature would be put into question. For example, the rotational zero of a distant neutron star (if a 2-D analog of such an object existed) would carry the earth's signature. All observers in the cosmos would feel the terrestrial privilege. Even more disturbing, the privilege would be personalized. Much as a relativistic frame is strictly speaking a one-man's-business (since the temperature-induced Brownian motion of the observer defines a frame of its own at every moment), a micro frame is observer-specific, too. However, unlike what holds good in macro relativity, all other observers would become an inalienable part of the first's spectacles, so to speak, since even memory provides no escape from a micro frame. Thus objective physics would single out one observer as "king" and turn all others into "slaves," so to speak, if the experiment worked out. It goes without saying that a prediction of this sort is almost infinitely unlikely to be confirmed.

In the same year in which Einstein published the claim that the facts of nature carry an observer-frame-specific signature, Ernst Kretschmer published his influential book *The Connections-Sensing Paranoia*.[14] Had he known about Einstein's subjectivistic physics, he might have been tempted to apply his book's title to it. *A fortiori*, the claim that objective physics contains the observer's signature, not only in his own frame but seemingly everywhere, appears like overstretching the belief in rationality.

Nevertheless time may be ripe for a "psychiatrization of physics," in the positive sense. Interface physics is close in spirit to psychoanalysis. Psychoanalysis claims that the "conscious" is not the true level but that there exists a more absolute level underneath which is the "subconscious." To search for signs of observer-centeredness in physics is analogous to Freud's proposal of searching for "*Fehlleistungen*" ("slips" of the tongue, etc.) that "give away" a deeper, if "censored," reality. A new method is therefore also available in physics. It might be called "physioanalysis" were it not for a better term already in existence (endophysics).

The price to pay is that "our beautiful physics" would be gone (Gert Eilenberger, personal communication, 1985). The idea of a "personalized cut" implicit in Bohr and Everett's micro relativity is indeed a blow to accepted dogma. Note, for example, that a manifest personalized cut would have the appalling implication that all records — including printed books — would be different if I had decided on a different rotational state a minute ago (to stick to the above unlikely example). The butterfly effect of chaos theory proposed by Ed Lorenz is equally counterfactual.[15] However, it could be accepted because it "only" affects the future, not the past.

If Everett's multiple-relative-states theory[12] and Gell-Mann's multiple-pasts theory[16] did not already exist, as accepted versions of the accepted theory of quantum mechanics, the above "multiple-records" theory, implicit in an artificial classical universe watched from within, would have been hard to communicate. The picture loses some of its strangeness, however, once one realizes that Bohr's endophysical idea of complementarity actually implies much more. Any "schismogenesis" (Gregory Bateson's neologism for an endogenous split) presupposes a larger unity. In the true physical reality, presupposed by Bohr, the quantum limit does not exist at all. Hence the content of all books might indeed be hallucinatory — it would still be fixed in the true reality (if only anyone could read it). Looked at against this backdrop, the multiplicity of "false readings," which seemed so startling only a moment ago, becomes a natural thing to expect. At the same time, the cabalistic idea of "two scriptures" — one with the merely human names in it and one with the "true names"[17] — is re-encountered in a profane context.

The repercussions of a positive outcome would go still farther, however. The "agent" which produces the content of the personalized interface at every moment would become a concept of physics — a reality worthy of empirical investigation. While the "dream-giving instance" of Descartes[18] has always been interpreted as a metaphysical agent in the West, the fact that the "assignment-giving instance" is rendered immanent by a positive outcome would detract also from the inaccessibility of the former.

In particular, "prison-exposing experiments" would become a paradigm in physics under the impact of a positive demonstration of observer-centeredness.[13] "Prison-busting experiments" would follow suit. The boundaries between physics, theology and gnosis, Eastern and Western thinking, would have to be redefined. The only way to cut this budding development out appears to be to find a counterargument to the proposal that a personalized cut can exist in an "exposable" fashion.

To conclude, a new aspect of endophysics has been stressed. The micro interface between observer and rest may be more than world-constituting — it may be subject to manifest choice. The "second causation" may prove to be manipulable. In the infinitely unlikely event that the third Copernican experiment would have a positive outcome, a "technology of world change" would become a realistic option.

Acknowledgments

I thank Achim Müller, Klaus Mainzer, Gregory Bateson, Jens Petersen, Jürgen Jonas, Bob Rosen, Hans Primas, Georg Franck, Julian Barbour, Siegfried Zielinski, Gerold Baier, Michael Klein, Marco Wehr, Harald Atmanspacher, Andreas Ammann, Klaus Zauner, David Köpf, Hans-Joachim Fuchs, Hanns Ruder, Vladimir Gontar, Horst Prehn, Ronald Imbihl, Karl Svozil, Stuart Kauffman, Kuni Kaneko, Ichiro Tsuda, Koichiro Matsuno, Kazuhiro Matsuo and Masao Yamaguti for discussions. For J. O. R.

References

1. S. Rozental, "The forties and the fifties," in *Niels Bohr — His Life and Work as Seen by His Friends and Colleagues* (Amsterdam, North-Holland, 1967), pp. 149–190.
2. O. E. Rössler, "Relative State Theory — Four New Aspects," *Chaos, Solitons & Fractals* **7**, 845–852 (1996).
3. K. Gödel, "An example of a new type of cosmological solution of Einstein's field equation of gravitation," *Rev. Mod. Phys.* **21**, 447 (1949).
4. R. J. Boscovich, "On space and time as they are recognized by us" ("*De spatio et tempore ut a nobis cognoscuntur*") (1755). English translation in: O. E. Rössler, "Boscovich covariance," in *Beyond Belief: Randomness, Prediction and Explanation in Science* (J. L. Casti and A. Karlqvist, eds.) (Boca Raton, CRC Press, 1991), pp. 69–87. Cf. Appendix to Chap. 10 of the present book.
5. A. J. Leggett, "Low temperature physics, superconductivity, superfluidity," in *The New Physics* (P. Davies, ed.) (New York: Cambridge University Press, 1989), pp. 268–288.
6. O. Avanel, P. Hakonen and E. Varoquaux, "Detection of the rotation of the earth with a superfluid gyrometer," *Phys. Rev. Lett.* **78**, 3602–3605 (1997).
7. H. Brodkey, *Die Taue sind gelöst* ("The ropes have been cut"), *Die Zeit*. Nr. 6, Feb. 2, 1996, p. 49.
8. *The Teaching of Buddha*, 696th edn. (Tokyo: Bukkyo Dendo Kyokai, 1991), p. 304.
9. E. Husserl, *Cartesian Meditations* (The Hague, Martin Nijhoff, 1960). (First French edn. 1929.)
10. V. Flusser, *Neue Wirklichkeit aus dem Computer* ("New reality out of the computer"), *gdi-impuls* **9** (4), 32–41 (1991).
11. O. E. Rössler and R. Rössler, "Is the mind–body interface microscopic?", *Theoretical Medicine* **14**, 153–163 (1993).
12. H. Everett III, "'Relative-state' formulation of quantum mechanics," *Rev. Mod. Phys.* **29**, 454–462 (1957).

13. O. E. Rössler, "Einstein completion of quantum mechanics made falsifiable," in *Entropy, Complexity and the Physics of Information* (W. H. Zurek, ed.) (Redwood City, Addison-Wesley, 1990), pp. 367–373.

14. E. Kretschmer, *Der sensitive Beziehungswahn* ("The Connections-Sensing Paranoia") (Tübingen, 1905).

15. E. N. Lorenz, *The Essence of Chaos* (Seattle: University of Washington Press, 1993), pp. 181–184.

16. M. Gell-Mann, *The Quark and the Jaguar* (San Francisco: Freeman, 1994).

17. N. Röller, *Migranten — Jabès, Nono, Cacciari* ("Migrants — Jabès, Nono, Cacciari") (Berlin: Merve-Verlag, 1995).

18. O. E. Rössler, "Endophysics — Descartes taken seriously," in *Inside Versus Outside* (A. Atmanspacher and G. J. Dalenoort, eds.) (Berlin: Springer-Verlag, 1994), pp. 153–161.

18
Limitology

Abstract

The three oldest limit statements are — perhaps — the "flaming sword" of the Bible, the notion of "Maya's veil" in India and Anaximander's claim that the Whole (One) is unrecognizable from within (only a cut is recognizable).

The first modern limit statement also takes the form of an interface statement. Boscovich's (1755) principle of the difference predicts that "the impressions generated" are invariant under certain objective changes of the universe (as when it "breathes" (increases and decreases in size) along with the observer and all forces. A century later, Maxwell (1872) recognized that a being who is part of a Hamiltonian universe cannot violate the second law of thermodynamics by performing micro observations followed by appropriate shepherding actions (like watching a gas of cold uranium atoms and setting barriers or removing them in an educated fashion using advanced modern gadgetry), whereas a being who is outside the same universe — a "demon" — can easily do the same thing. He thereby was able to predict that micro motions will in general be "impalpable" (p. 154 of his *Theory of Heat*, 1872). After this prediction had come true, half a century later, Bohr (1927) claimed that his own "complementarity principle" marks the limit of what an internal observer of a classical-continuous world can hope to measure. Four years later, Gödel (1931) discovered his discrete limit (an inaccessibility in finitely many steps of certain implications of a formal system from within). Unlike Bohr's intutive principle, Gödel's is a hard theorem. However, Bohr's also is harder in a sense because it asserted a contradiction between the inside and the outside view, while Gödel only found that a separating boundary existed between reachable and unreachable truths. Bohr's limit would be a "distortion" limit, Gödel's an "inaccessibility" limit.

Distortion limits were subsequently also found by Ed Moore in 1956 (an analog to uncertainty in certain dissipative automata) and Donald Mackay in 1957 (irrefutability of certain false statements, like the attribution of free will, to deterministic automata). The subsequent limits of Bob Rosen in 1958 (to self-reproduction in category theory), Rolf Landauer in 1961 (to dissipation-free computation) and Ed Lorenz in 1964 (butterfly effect in chaotic dynamics) appear to be inaccessibility limits again. The same holds good for the finite combinatorial inaccessibility (NP-completeness) discovered in the 1970's. The most recent limit of Yukio Gunji (inconsistency of language games) appears to be a distortion limit again.

Do these findings justify the introduction of a new discipline? Two points are in favor of the idea. First, new limits become definable in the fold of the new paradigm. For example, "ultraperspective" (which is inaccessible to animals) plays a fundamental role in both mathematical economics and science. The very notion of limits implies the adoption of two perspectives simultaneously. Second, any distortion limit "splits" the single (exo) reality into many different internally valid (endo) realities. Some of the objective features become interface-specific. This allows a new question to be asked: Can the "mirage properties" implicit in a distortion limit always be identified ("tagged") from the inside, even though they of course cannot be removed by definition? If so, relativity may turn out to be a first example in physics. Related observer-objective phenomena should then occur and be identifiable on the micro level (micro relativity). The famous Goldstein–Kerner "no interaction theorem" (which precludes a frame-dependent description of gases) would acquire a deep significance. Two previously incompatible fundamental theories of physics, namely relativity and quantum mechanics, would be unexpectedly unified by limitology.

A third point is in favor of limitology: distortion limits always leave a loophole. They exist because an objective picture (which no longer depends on the observer) can only be obtained by making the observer explicit — a feat impossible to accomplish from the inside. Nevertheless the impossible can be achieved — on the modeling level. The "artificial universe approach" to the real world therefore qualifies as a new type of measurement. At the same time the computer acquires an unexpected fundamental role.

Acknowledgments

I thank John Casti, Joe Traub, Oliver Morton, Rolf Landauer, Atlee Jackson, Patrick Suppes, Greg Chaitin, Bob Rosen, Chico Doria, John Horgan, Piet Hut, Roger Shepard, Mel Cohen, Hao Bai-Lin, Karl Svozil, Brian Artur, Artur Schmidt and Yukio Gunji for discussions. For J. O. R.

19
Our Rainbow World

Coauthor: Peter Weibel

Summary

Reality can be divided up into two parts — rainbow-like and non-rainbow-like. The rainbow variety has largely been uninvestigated so far. A rainbow is a virtual image which pretends to be a real object. Such "chimeras" were a theoretical curiosity in the past. In the modern age of virtual reality, they have become a technological challenge. At the same time a number of phenomena of the real world — ranging from the very small to the very fleeting — may turn out to belong to the alternative class. The manipulation of rainbow phenomena is possible in principle. An "alien" type of technology will be up to the task.

19.1 Introduction

A rainbow's two ends stand in a pot of gold, as is commonly known. Its location, however, provides sort of a problem since a rainbow presents itself differently to every observer. Although the rainbow is really a virtual image of the sun, as Descartes demonstrated, it resembles a real object. Could it be that other "real" objects belong to the same class?

19.2 An Old Question Revisited

To what extent does objective reality depend on the observer? Since the invention of perspective during the Renaissance, and the invention of group theory (Helmholtz–Lie groups) in the 19th century, we have known that the appearance of the world depends in a lawful way on the location of the observer. Accordingly, computer programs of the "virtual reality" type generate — out of an absolute (invariant) representation stored in the computer's memory — a "lawfully noninvariant" (that is, covariant) representation. Although the lawful distortions of perspective vision are major and tantalizing if one pays attention to them, they are inconspicuous when they occur and rather convey to us a secure feeling of an invariant "objective" reality existing out there which we have firmly in our grips.

The rainbow renders this security brittle. Virtual reality programs containing rainbows have yet to be developed. The transformation rules for such objects are different from those needed for ordinary objects. The reason has to do with the fact that a rainbow is a rather special entity. It is a distorted virtual image of the sun. This means that when the observer is traveling — for example with a speeding car — so is the rainbow. And if the observer increases the distance between his or her two eyes with the aid of field glasses or a pair of mirrors (a situation which can be mimicked in a virtual reality simulation by increasing the size of the internal representation of the observer), the rainbow does not shrink but retains its size and infinite distance — despite the fact that it is "overlaid" on objects in the foreground (like mountains, forests and villages). While no one doubts that a pot of gold waits at the foot, it turns out to be somewhat difficult to simultaneously stay and watch and sneak toward the right place to catch the trophy.

Thus the properties of certain objects — in the present case their size and their location — depend on properties of the observer (like her or his size and location) in a way which transcends the familar distortions of 3-D Helmholtzian perspective and 4-D Einsteinian relativity. Could it be that the principle illustrated by the rainbow is of a broader significance?

19.3 The Interface Between Observer and Environment

Being a part of the universe, the observer cannot see the world from an objective vantage point. The "homogeneous matrix algorithm" of the flight simulators of virtual reality (cf. Ref. 1) shows how nontrivial a task it is to generate the right user interface. The observer is (as mentioned) not only not intimidated by the rich and changing structure of the many sequentially applying perspectives generated by his own motion but thrives on this kind of packaging in order to effortlessly arrive at the correct invariant mental image. "The intimacy of a head near one's own is like the lights and doorway of a house."[2]

In principle, there are more parameters waiting to be investigated than just the observer's location and size. Motion of the observer comes to mind immediately. It evokes both "visual flow" phenomena (Gibson) and relativistic distortions which can indeed be reproduced in simulation.[3,4] Next, one thinks of recurrent motions of the observer, like a shaking of the head. The effects on the interface can be dramatic — especially if the shaking is fast. Irreparable damage can be done to the goal of finding an invariant representation under such a predicament.

Historically, the interface problem was apparently first seen by Boscovich in 1755.[5,6] He raised the question of what might be expected to happen if the observer and the surrounding world were shrunk concomitantly, along with all forces, in case matter is based on point particles only (as he had first proposed). Obviously, "the same impressions will be generated in the mind." That is, the interface remains unaffected. Similarly, nothing would change for the obsever if this shaking of the head were accompanied by a matching shaking of the rest of the universe along with all forces. The time-dependent features of the interface, first glimpsed by Boscovich, therefore deserve a closer look.

19.4   A Brownian Observer's Interface

Brownian motion (like that of an ink particle in water observed with a microscope) is interesting in its own right because of the counterintuitive energy and momentum conservation involved. A frictionless system of springs and balls shaking chaotically all over is the simplest example. Every particle has relative to every other, and every collection of others, the same mean kinetic energy. An observer who is built up from particles that are in random thermal motion is another case in point. He stands in a nontrivial dynamical relationship with the rest of the universe. Archimedes first saw that the "center of mass" can never be moved from within the world. (His famous saying "Give me where I might stand and I shall move the Earth" belongs in this context.) How does the rest of the universe appear to an internally moving observer? This question can meaningfully be asked only in the age of fast computers, since the simulation of many particles simultaneously is an option which was not available in the past.[8]

Every external object will then be found to perform a Brownian motion relative to the observer, by the observer. The strength of its Brownian motion depends on the object's mass: the smaller the mass, the larger the apparent thermal agitation. This is because the center of mass of the observer and the center of mass of the external object are linked by a center-of-mass-preserving relative Brownian motion. A very-small-mass object can therefore never be observed in an error-free manner by a thermally shaken observer. The thermal noise of the observer "infects" the object in such a way that it is the latter which appears to be thermally agitated by the observer's temperature — even though, in reality, the object may be motionless at zero temperature.

The effect thus is the same as if the observer were a "Brownian particle" himself (or herself). "How does the world look like to a particle in Brownian motion?" is, therefore, a meaningful question to ask. The virtual reality paradigm can be used to obtain the answer.

19.5   A Quaker's World

Finding the right tranquility of mind to look at the situation in the right way is not easy. Numerically, too, the task is demanding. It consists in designing a whole reversible universe in the computer. The internal "eye" (that is, the internal observer) must be built up from the same reversibly moving micro constituents as the rest of the artificial universe. The special thermal (momentum-conserving) relationship between the "eye particle" and a particular particle in the same universe can then be watched by a human "superobserver" who sits outside the simulated universe wearing the right kind of goggles.

It will be rewarding to implement this scenario by the year 2010, say. A fleeting glance at some of the unusual rainbow phenomena which are expected to emanate from the contraption can, however, already be risked.

19.6   False Uncertainty

An irreducible uncertainty arises — as a first implication. The chaos inside the observer "translates" into a chaos outside the observer. Apart from the unit thermal noise energy in the observer (E), there is a second intrinsic constant (T) involved. It is a characteristic time interval which is related to the mean "rearrangement interval" (of two neighboring particles passing by each other) valid in the observer. After this time interval has passed, the whole micro dynamics inside the observer changes its course relative to the external object. A precise calculation of T for classical billiard systems still constitutes an open problem.[9] What is already clear, however, is that, if the observer is alone in his universe with the micro object, the center of mass of the latter performs a Brownian motion relative to the observer in such a way that the observer reverses course, every unit time interval T, relative to the external object and vice versa.

The resulting "relative diffusion" between external object and observer is governed by the product of E and T, divided by the object's mass (M). This result holds good not only if the external object is "directly" coupled to the observer (Chap. 6) but also in the more general case of "indirect" coupling (via a measuring chain). For even an amplification type measuring chain cannot "undo" the objectively existing mutual relationship between observer and object.

The resulting "uncertainty" shows some similarities to quantum mechanics.[10,11] This is because the presence of a "diffusion law" of the type described above (an action, like E times T, divided by a mass, like M) suffices to generate the Schrödinger equation[12,13] which governs quantum mechanics.

19.7 False Certainty

We still need to find out what happens when the observer tries to force a micro object into a certain definitive observational state. For example, the measuring situation could be arranged in such a way that the micro object is bound to reveal its position in a "yes-or-no decision" at some point in time. The task in hand is analogous to the measurement of an "eigen state" in quantum mechanics. Such a restrictive type of measurement can certainly be performed in our simulated world by arranging many sensors around the objects.

A new phenomenon arises in this case. While the previous finding (uncertainty) did not yet qualify as a rainbow phenomenon in the strict sense — since a mere "blurring" does not yet bring along a new phenomenological quality — in the present case a new quality does arise. It is the quality that a well-defined localization in position space (or momentum space, respectively) emerges for the observer — even though that location is at variance with the correct location valid from the outside. This falsity is bound to occur because if the observed location of the object were identical with the correct location, the relative Brownian motion of the object with respect to the observer would have been eliminated in effect (even though this cannot happen, as we saw). Hence the apparent location of the object, valid in the interface, must in general be different from the objectively valid location.

This prediction can be tested in the proposed computer simulation. Since everything which happens in the simulation is known explicitly, it is possible to compare the content of the computed interface with what really happens to the particle in question. This comparison will, of course, be a privilege reserved for the external superobserver since the internal observer cannot leave the interface.

19.8 A Lesson for Superobservers

The yes-or-no decision which makes its apparition in the interface is different from the correct result because it depends not only on the object's own dynamics but also on the internal dynamics of the observer. This explains why the measured "eigen state" differs from the true state. The same result — a perturbed measurement — is valid in Nelsonian stochastic mechanics.

However, in the artificial universe an unexpected complication arises. In Nelson's diffusion theory, the quantum decisions ("eigen states") are permanent. Here, the distortion of the objective world is such that the recorded state — as it arises on the interface — does not cease to depend on the momentary state of motion of all particles inside the observer. The distortion therefore goes on indefinitely!

The interface is a momentary state of affairs. All measurements — no matter how long the measuring chain may be in space and time — are determined by the momentary relationship between the internal dynamics of the observer and the dynamics of the rest of the universe. Therefore, the external superobserver, who watches the momentary interface world as a function of time, necessarily records a "superposition" (an integral) over all the different successive eigen worlds (rainbow worlds).

This unexpected predicament rings a bell in the context of quantum mechanics. It reminds one of the so-called "measurement problem." The latter consists in the fact that the wave function discovered by Schrödinger is continuous and smooth (describing likelihoods of finding a certain measurement result) but that actual results are not. The problem is usually solved in an ad hoc fashion: one postulates that the different possible measurement-generated eigen worlds which could occur at each moment are "shielded" from each other so that only one is allowed to be manifest in general (or for the observer, respectively). The latter alternative is adopted in the case of Everett's "relative state" theory (where the other worlds are real relative to some other state of the observer[14]). The former alternative is the essence of the usual so-called "reduction of the wave packet" interpretation (cf. Ref. 15 for a collection of the relevant literature including Everett's paper).

In addition, there exists in quantum mechanics a third interpretation, due to Bell,[16] in which the different eigen worlds do not exist permanently. Each world is here confined to a very short time window. Both Everett's and Bell's interpretation are generally considered outlandish (cf. Chap. 13). In the present context of the predicament incurred by the superobserver of a classical world (a "blur" among all the lower level eigen worlds), Bell's proposal — mutual hermeticity of short-lived whole worlds — takes on the role of the proverbial "egg of Columbus."

The external view and the internal view are suddenly reconciled. The internal observer does not "register" being placed into a different quantum world from one moment to the next. This is not a postulate but a result. For "worlds" by definition are complete. They cannot contain a residue from another world. This is the missing piece of information that needs to be given to the superobserver.

The "superposition" of interface-specific worlds, experienced by the superobserver, turns out to be an artifact. For if the superobserver participated in the interface, being unable to escape from it with the aid of his outside memory, the phenomenon of superposition would simply disappear. At every moment, a single consistent "eigen world" would be applicable, complete with its own recorded past and anticipated future. This makes the job of a demiurge — if he wants to register the implications his own actions (laws and initial conditions) entail for the inhabitants — unexpectedly hard.

19.9   A Rainbow Movie

The distortion of a universe mirrored in an internal interface suggests the term "rainbow world." In the one world, Schrödinger's cat is alive and well, in the other the "hellish contraption"[17] has taken the other course. If the branching has taken place some time ago already, the one world will harbor a roaming frisky cat while the other is marred by an advanced stage of organic decomposition. The two rainbow worlds would be equally hallucinatory, generated by the same exo reality. The "rainbow movie" of alternating worlds at the same time reminds one of real life, which also can look quite different from moment to moment — although here the bridge of memory is of course available.

One feature of the rainbow movie is not at all different from everyday experience, however: the fact that to each moment there belongs a world ("eigen world") and vice versa.[18] This "binding" is, unlike the "jumps" that go with it, recognizable also from the inside of the model universe. It is bound to cause wonder among the inhabitants. The fact that they are glued to a privileged moment in time called "the world as it is real now" will stand out to them as being at variance with a science which lacks the concept of a privileged now.

Therefore, the existence of a shining now might become a focus of attention also in our own world. Now and rainbow would touch each other in a pot of gold.

19.10   Conclusions

The virtual reality paradigm lends scientific credence to the topic of the "interface."[19.20] The momentary position of the "index glove" inside a simulated world distorts that world in a way which only makes it palpable as an invariant reality. The generation of such artificial interfaces is not easy and requires fast computers, as is well known. Experiments with this kind of interface are a current challenge. For example, how does a rainbow look when programmed into a virtual reality? And: How does it look when it is viewed through the vertical pupil of a cat? Or through a 100-meter-long pupil?

This program then leads over to temporally changing realities. What if the position of the "eye" and of that of an external object are correlated? Will such correlated objects tend to disappear in the perceived interface?

A third type of experiments will involve reversible simulated worlds. This proposal includes the study of the predicament of an ice-skater who cannot get rid of his whole-body angular momentum unless he pirouettes. At still greater degrees of sophistication, the proposal will include an Archimedean system of balls and springs — like a giant model drug molecule — to play with. Eventually, as the computers grow faster, a whole ice-skater will come into sight again — but this time watched completely from the Archimedean frog's perspective of total reversibility. The first detailed report about the properties of such a "conservative virtual reality" will come in in the year 2010, if everything goes well.

In this way, a new "hopeful suspicion" could be arrived at: the virtual reality paradigm may reveal more about our own world than the usual course of science has led us to believe. Is the whole world a rainbow world? Only after such a suspicion has taken hold do new "diagnostic tools" — to confirm and to manipulate it — have a chance to be developed seriously.

To conclude, the rainbow has not lost any of its childhood magic. To simulate it interactively, an advanced type of VR is required. At the same time, a new attitude is fostered — the walls which surround the sparkling VR-like now at every moment acquire a new tangibility. Does the hermetic paradigm of computer-generated worlds perhaps hold the key to making our own world less hermetic? The rainbow would become a door in the sky.

Acknowledgments

We thank Roland Fischer, Jerzy Gorecki, Jim Crutchfield, Rob Shaw, Michael Klein, Marcus Fix and Sven Sahle for discussions. For J. O. R.

References

1. W. M. Newman and R. F. Sproull, *Principles of Interactive Computer Graphics*, 2nd edn. (McGraw-Hill, New York, 1979).

2. T. Rodney, *The Runaway Soul* (Farrar, Straus and Giroux, New York, 1991).

3. I. E. Sutherland, "Computer inputs and outputs," *Scientific American*, 1966. Sep. issue.

4. I. E. Sutherland, "Computer displays," *Scientific American*, 1970, June issue.

5. R. J. Boscovich, *De spatio et tempore, ut a nobis cognoscuntur* ("On space and time, as they are recognized by us"), in: J. M. Child (ed.), R. J. Boscovich, *Theory of Natural Philosophy*, Latin-English edition (Open Court, Chicago, 1922), pp. 404–409. Reprint of the English translation (MIT Press, Cambridge, Mass., 1966), pp. 203–205. For a modern retranslation see Chap. 10 of this book, and Ref. 7 below.

6. O. E. Rössler, "Boscovich covariance," in: J. L. Casti, A. Karlqvist (eds.), *Beyond Belief: Randomness, Prediction and Explanation in Science* (CRC Press, Boca Raton, 1991), pp. 69–87. Cf. also Chap. 10 of the present book.

7. R. Fischer, "A neurobiological re-interpretation and verification of Boscovich covariance, postulated in 1758," *Cybernetica* **34**, 95–101 (1991).

8. B. J. Alder and T. E. Wainwright, "Phase transitions for a hard-sphere system," *J. Chem. Phys.* **27**, 1208 (1957).

9. O. E. Rössler, "Four open problems in four dimensions," in: G. Baier and M. Klein (eds.), *A Chaotic Hierarchy* (World Scientific, Singapore, 1991), pp. 365–369.

10. O. E. Rössler, "Endophysics," in: J. L. Casti and A. Karlqvist (eds.), *Real Brains, Artificial Minds* (North-Holland, New York, 1987), pp. 25–46. Cf. Chap. 6 of the present book.

11. O. E. Rössler, "A possible explanation of quantum mechanics," in *Advances in Information Systems Research* (G. E. Lasker, T. Koizumi and J. Pohl, eds.) (International Institute for Advanced Studies in Systems Research and Cybernetics, Windsor, 1991), pp. 581–589.

12. I. Fényes, "A probability-theoretical explanation and interpretation of quantum mechanics" (in German), *Z. Phys.* **132**, 81–106 (1952).

13. E. Nelson, "Derivation of the Schrödinger equation from Newtonian mechanics," *Phys. Rev.* **150**, 1079–1085 (1966).

14. H. Everett, III, "'Relative-state formulation' of quantum mechanics," *Rev. Mod. Phys.* **29**, 454–462 (1957).

15. J. A. Wheeler and W. H. Zurek, *Quantum Theory and Measurement* (Princeton University Press, Princeton, 1983).

16. J. S. Bell, "Quantum mechanics for cosmologists," in: C. J. Isham, R. Penrose and D. Sciama (eds.), *Quantum Gravity 2* (Oxford, Clarendon, 1981), pp. 611–637. Reprinted in his *Speakable and Unspeakable in Quantum Mechanics* (Cambridge University Press, Cambridge, 1987), pp. 117–138.

17. E. Schrödinger, "The present situation in quantum mechanics" (in German), *Naturwissenschaften* **23**, 807–812; 823–828; 844–849 (1935). English translation: *Proceedings of the American Philosophical Society* **124**, 323–338 (1980).

18. D. Deutsch, "Three connections between Everett's interpretation and experiment," in: R. Penrose and C. J. Isham (eds.), *Quantum Concepts in Space and Time* (Clarendon, Oxford, 1986), pp. 215–225.

19. P. Weibel (ed.), *Ars Electronica 1988, Philosophies of the New Technology* (Merve-Verlag, Berlin, 1986).
20. P. Weibel, "Virtual worlds — the emperor's new bodies," in *Ars electronica 90*, Vol. 2 (G. Hattinger, M. Russel, C. Schöpf and P. Weibel, eds.) (Veritas-Verlag, Linz, 1990), pp. 9–38.

20
The Future of Endophysics

What do yin-and-yang and Ed Lorenz's butterfly (whose wing flap in Brazil leads to a tornado in Texas half a year later) have in common? It is the outside perspective.

Only when going very far outside — beyond the end of the world — do certain aspects of the world begin to make sense. Lorenz, for example, would need two runs of the world in order to demonstrate the butterfly effect — something which can never be achieved from the inside. However, the computer makes it possible today to understand this.

How does the world look like if you are an internal part of it? Is the world perhaps only a "difference," as Boscovich claimed? An interface? An effective forcing function (to use the correct term from engineering science)?

It is in principle possible to identify those features of the world which exist only from the inside. Both quantum mechanics and relativity may be virtual realities. The computer paradigm of molecular dynamics simulation, which is pure chaos, enables the investigation of explicit interfaces — and thereby suggests a novel interpretation of both relativity and quantum mechanics.

The new concept of interface reality also implies a new option. The micro interface — if it indeed exists — can presumably be manipulated. Micro relativity — the science of the microscopically exact interface — would then lead to a new technology. If the world, valid at this now, is interface-generated, both the world and the now can be changed in principle — by "merely" manipulating the interface.

In 1949, Gödel found a time machine based on the weakly nonlinear Einstein equations of general relativity — the famous "Gödel solution." The strongly nonlinear interface of micro relativity could in principle lead to even more astounding effects. The walls of the Cartesian prison called the world can perhaps be bent open a little bit. This is the hope of endophysics.

Acknowledgments

I thank H. T. Leong and R. Duhlev for invaluable encouragement, help and co-operation. Thanks go also to Peter Weibel (for the 1992 "ars electronica"), Klaus-Peter Zanner (for "lampsacus") and the "Käfer" (for illuminating our lives). For J. O. R.

Sources

0. Preface by Peter Weibel
   Translated and adapted from: *Vorwort*, in: O. E. Rössler, *Endophysik —
   die Welt des inneren Beobachters* (P. Weibel, ed.), pp. 9–12 (Merve-Verlag,
   Berlin, 1992).

1. Anaxagoras' Idea of the Infinitely Exact Chaos
   Adapted from: "Anaxagoras' idea of the infinitely exact chaos,' in *Teaching Nonlinear Phenomena, Vol. II — Chaos in Education* (G. Marx, ed.),
   pp. 99–113 (National Center for Educational Technology Publications,
   Veszprém, 1987), and from: *Anaxagoras'Idee des unendlich exakten Chaos*
   (in *Endophysik*).

2. How Chaotic Is the Universe?
   Adapted from: "How chaotic is the universe?", in *Chaos* (A. V. Holden,
   ed.), pp. 315–320 (Manchester University Press, Manchester, 1986), and
   from: *Wie chaotisch ist das Universum?* (in *Endophysik*).

3. A Possible Explanation of Quantum Mechanics (abstract)
   Abridged and adapted from: "A possible explanation of quantum mechanics," in *Advances in Information Systems Research* (G. E. Lasker, T.
   Koizumi and J. Pohl, eds.), pp. 581–589 (International Institute for Advanced Studies in Systems Research and Cybernetics, Windsor, 1991), and
   from: *Eine mögliche Erklärung der Quantenmechanik* (in *Endophysik*).

4. Endophysics — Physics from Within (abstract)
   Adapted from: "Endophysics — physics from within,' in *International
   Conference on Cybernetics* (P. Kugler, org.) (Amsterdam, Aug. 1991),
   and from: *Endophysik — Physik von innen* (in *Endophysik*).

5. Letter by David Finkelstein, June 23, 1983, originally published in facsimile
   in *Endophysik*. Republished with the permission of the sender.

6. Endophysics
   Adapted from: "Endophysics," in *Real Brains, Artificial Minds* (J. L. Casti
   and A. Karlqvist, eds.), pp. 25–46 (North-Holland, New York, 1987),
   and from: *Endophysik* (in *Endophysik*). Cf. also the Japanese edition of
   Real Brains, Artificial Minds (M. Nakamura, transl.), pp. 29–55 (Kyoritsu,
   Tokyo, 1991).

7. The Two Levels of Reality — "Exo" and "Endo' (abstract, coauthor P.
   Weibel)
   Adapted from: "The two levels of reality — exo and endo," in *Artlab
   1st Symposium — The Current Condition and the Future of Digital Art*

(Canon, Tokyo, June 29, 1991), p. 16, and from: *Die zwei Ebenen der Realität — "Exo" und "Endo"* (in *Endophysik*).

8. Explicit Observers
   Adapted from: "Explicit observers," in *Optimal Structures in Heterogeneous Reaction Systems* (P. J. Plath, ed.), pp. 123–138 (Springer-Verlag, Berlin, 1989), and from: *Explizite Beobachter* (in *Endophysik*).

9. The Endo Approach (abstract, coauthor J. Rössler)
   Adapted from: "The endo approach," letter of homage to Peter Weibel (November 23, 1990), and from: *Der Endo-Zugang* (in *Endophysik*).

10. Boscovich Covariance
    Adapted from: "Boscovich covariance," in *Beyond Belief — Randomness, Prediction and Explanation in Science* (J. L. Casti and A. Karlqvist, eds.), pp. 69–87 (CRC, Boca Raton 1991, and from: *Boscovich-Kovarianz* (in *Endophysik*).

11. Can There Be Two Chains of Causality? (abstract)
    Adapted from: "Can there be two chains of causality?" (July 20, 1991), and from: *Kann es zwei Kausalketten geben?* (in *Endophysik*).

12. The Golden Thread of Paradise — An Implication of the Causally Interpreted Bell–Everett Theory (abstract)
    Adapted from: "The golden thread of paradise — an implication of the causal Everett–Bell theory," letter to Joseph Ford (June 24, 1991), and from: *Der goldene Faden des Paradieses — Eine Implikation der exokausal interpretierten Everett–Bell-Theorie* (in *Endophysik*).

13. Into the Same Rivers We Step and Step Not, We Are the Same and We Are Not: On the Origin of the Now
    Adapted from: *In dieselben Flüsse steigen wir und steigen wir nicht, wir sind dieselben und wir sind es nicht — Über den Ursprung des Jetzt*, in *Chaos und Pädagogik* (J. R. Bloch, ed.), unpublished, and from: *In dieselben ...* (in *Endophysik*).

14. A Possible Explanation of Spin (abstract)
    (August 8, 1996.)

15. Microconstructivism
    Adapted and translated from: *Mikro-Konstruktivismus*, in *Lab — Jahrbuch 1995-96 für Künste und Apparate der Kunsthochschule für Medien Köln* (U. Reck, S. Zielinski, N. Röller and W. Ernst, eds.), pp. 208–227 (Verlag der Buchhandlung Walther, Cologne, 1996), and from: "Micro constructivism," *Physica* **D75**, 438–448 (1994).

16. Interference Is Exophysically Absent (abstract, coauthor M. El Naschie)
    Abridged and adapted from: "Interference through causality vacillation"
    (with M. S. El Naschie), in *Symposium on the Foundations of Modern
    Physics 1994* (13–16 June 1994, Helsinki, Finland), Extended Abstracts,
    Report Series Publication of the Department of Physical Sciences, University
    of Turku, 1994.
17. Second Causation — Assignment Conditions
    Adapted from: "Second...," in *Causality, Self-Organization and Computability: Interdisciplinary Approaches of Physics, Chemistry, Biology
    and Philosophy*, proceedings of a conference held in Bielefeld in 1995 (A.
    Müller and K. Mainzer, eds.), to appear.
18. Limitology (abstract)
    Adapted from: "Limitology," in *On Limits* (J. L. Casti and J. Traub, eds.),
    p. 17. Abstracts of the Santa Fe Institute Workshop "Limits to Scientific
    Knowledge," May 24–26, 1994, Santa Fe Institute Publication 94-10-056.
19. Our Rainbow World (coauthor P. Weibel)
    Adapted from: "Our rainbow world," in *Ars electronica 92: The World
    from Within — Endo and Nano* (K. Gerbel and P. Weibel, eds.), pp. 13–21
    (PVS Verleger, Linz, 1992).
20. The Future of Endophysics
    (September 21, 1996.)

Other Works on Endophysics
by the Author

1. The "Glass Cutter's Principle" — Glass Cannot Be Used to Cut Glass
   (Letter to C. F. von Weizsäcker, Sep. 1966.)
2. Chaos and Chemistry (1981)
   In *Nonlinear Phenomena in Chemical Dynamics* (C. Vidal and A. Pacault,
   eds.), pp. 79 87 (Springer-Verlag, Berlin, 1981).
3. Macroscopic Behavior in a Simple Chaotic Hamiltonian System (1982)
   In *Dynamical Systems and Chaos* (L. Garrido, ed.), *Lecture Notes in
   Physics* **179**, 67 77 (1983).
4. Classical Quantization — Two Possible Approaches (1983)
   In *Chaotic Behavior in Quantum Systems: Theory and Applications* (G.
   Casati, ed.), pp. 345 351 (Plenum, New York, 1985).
5. An Estimate of Planck's Constant (1984)
   In *Dynamical Phenomena in Neurochemistry — Theoretical Aspects* (P.
   Erdi, ed.), pp. 16-18 (Publications of the Institute of Theoretical Physics
   of the Hungarian Academy of Sciences, Budapest, 1985).
6. Indistinguishability Implies Quantization (1985)
   In *Collected Papers Dedicated to Professor Kazuhisa Tomita on the Occasion of his Retirement from Kyoto University* (S. Takeno, T. Kawasaki
   and H. Tomita, eds.), pp. 280 288 (Publication Office of *Progress of Theoretical Physics*, Kyoto, 1987).
7. A Possible Explanation of Quantum Mechanics (1985)
   In *Advances in Information Systems Research* (G. E. Lasker, T. Koizumi
   and J. Pohl, eds.), pp. 581 589 (Windsor: The International Institute for
   Advanced Studies in Systems Research and Cybernetics, 1991); cf. also
   the editorial by G. Eilenberger, *Phys. Bl.* **41**, 417 (1985).
8. A Chaotic 1-D Gas — Some Implications (1986)
   In *The Physics of Phase Space* (Y. S. Kim and W. W. Zachary, eds.),
   Lecture Notes in Physics **278**, 9-11 (1987).
9. Quasiperiodization in Classical Hyperchaos (with M. Hoffmann, 1986)
   J. Comput. Chem. **8**, 510–515 (1987).
10. Explicit Dissipative Structures (1986)
    In *Invited Papers Dedicated to Ilya Prigogine on the Occasion of His 70th
    Birthday* (A. van der Merwe and G. Nicolis, eds.), *Found. Phys.* **17**,
    679 688 (1987).
11. Symmetry-Induced Disappearance of Reality — The Leibniz Effect (1987)

In *Art and the New Biology: Biological Forms and Patterns* (P. Erdi and L. Bornstein, eds.), *Leonrado* **22**, 55–59 (1989); cf. also: *Der Leibniz-Effekt: Das symmetrie-induzierte Verschwinden von Realität*, in *Strategien des Scheins — Kunst, Computer, Medien* (F. Rötzer and P. Weibel, eds.), pp. 277–289 (Munich: Boer, 1991).

12. Four Open Problems in Four Dimensions (1989)

    In *A Chaotic Hierarchy — Festschrift to O. E. Rössler on the Occasion of His 50th Birthday* (G. Baier and M. Klein, eds.), pp. 365–369 (Singapore: World Scientific, 1991).

13. Boscovich's Observer-Centered Explanation of the Nonclassical Nature of Reality (1990)

    In *Symposium on the Foundations of Modern Physics 1990* (13–17 Aug. 1990, Joensuu, Finland), Extended Abstracts, pp. 153–156. Report Series, Publications of the Department of Physical Sciences, University of Turku, 1990.

14. How Many "Demons" Do We Need? Endophysical Self-Creation of Material Structures and the Exophysical Mastery of Universal Libraries (with G. Kampis, 1990)

    In *Cybernetics and Systems 90* (R. Trappl, ed.), pp. 27–34 (Singapore: World Scientific, 1990).

15. The Future of Chaos (1990)

    In *Applied Chaos* (J. H. Kim and J. Stringer, eds.), pp. 457–465 (New York: Wiley, 1992).

16. Bell's Symmetry (1990)

    Symmetry: Culture & Science, submitted.

17. Measuring the Future (1991)

    In *Measurement and Selfsimilarity*, Proceedings of the 17–24, Feb. 1990 Zeinisjoch Conference (P. J. Plath, ed.), submitted.

18. The Endo Approach (with J. O. Rössler, 1990)

    Applied Mathematics and Computation **56**, 281–287 (1993).

19. *Ist unsere Welt eine virtuelle Realität?* [Is Our World a Virtual Reality? (1991)]

    In *Cyberspace — Zum medialen Gesamtkunstwerk* (F. Rötzer and P. Weibel, eds.), pp. 256–260 (Munich: Boer, 1993).

20. Why Science?: The Anti-Lovecraft Effect (1991)

    Janna-Fix-Paper-Symposium (M. Fix, ed.) (Internet Publication, Institute for New Media, Frankfurt, 1995).

21. *Chaos und Endophysik* ["Chaos and Endophysics" (1991)]

[In *Quanten, Chaos und Dämonen* (K. Mainzer and W. Schirmacher, eds.), pp. 220–235 (Mannheim: B. I. Wissenschafts-Verlag, 1994).

22. Privileged Frames — From Endo Covariance to Exo Invariance (with P. Weibel, 1992)
    In *Endophysics — The World from Within — A New Approach to the Observer Problem with Applications in Physics, Biology and Mathematics* (G. Kampis and P. Weibel, eds.), pp. 235–238 (Santa Cruz: Aerial, 1994).

23. The Brain as a Measuring Apparatus (with J. Parisi, 1992)
    In *Endophysics — The World from Within, a New Approach to the Observer Problem with Applications in Physics, Biology and Mathematics* (G. Kampis and P. Weibel, eds.) (Santa Cruz: Aerial), submitted.

24. Some Remarks on the Experimental Realization of a Mind Machine (with J. Parisi, 1992)
    BioSystems **42**, 207–208 (1997).

25. The Transparent Medium (1992)
    In *On Justifying the Hypothetical Nature of Art and the Nonidentity Within the Objective World* (P. Weibel, T. Grunert and M. Janssen, eds.), pp. 121–126 (Cologne: Verlag der Buchhandlung Walther König, 1992).

26. The Shadow Principle — Is Physical Reality Full of Correlations with the Observer? (1992)
    In *International Symposium on Information Physics* (I. Tsuda, B. A. Huberman, O. E. Rössler and I. Percival, eds.), pp. 134–141 (Iizuka, Fukuoka, Fujiki Publishing Co., 1992).

27. *Endophysik, eine neue kopernikanische Revolution?* ("Endophysics, a new Copernican turnabout?", with F. Rötzer, 1992)
    Kunstforum **124**, 212–220 (1993).

28. A New Principle for Identity Generation: Three Equal Particles on the Ring — A Kaleidoscope of Classical Exchange Symmetry (1992)
    In *Austria: Biennale di Venezia 1993 — Representatives* (P. Weibel, ed.), pp. 257–272 (Cologne: Verlag der Buchhandlung Walther König, 1993).

29. Is the Mind–Body Interface Microscopic? (with R. Rössler, 1992)
    Theoretical Medicine **14**, 153–165 (1993).

30. Endophysics — Descartes Taken Seriously (1993)
    In *Inside Versus Outside: Endo- and Exo-Concepts of Observation and Knowledge in Physics, Philosophy and Cognitive Science* (H. Atmanspacher and G. J. Dalenoort, eds.), pp. 153–161 (Berlin: Springer-Verlag, 1994).

31. *Chaos, Rationalität und das "bad-trip"-Problem* ("Chaos, Rationality and the Bad-Trip Problem," 1993)

In *Evolution, Entwicklung und Organisation in der Natur* (V. Braitenberg and I. Hosp, eds.), pp. 107–119 (Reinbek: Rowohlt, 1994).

32. Interference Through Causality Vacillation (with M. S. El Naschie, 1994)
In *Symposium on the Foundations of Modern Physics 94* (13–16 June, 1994, Helsinki, Finland), Extended Abstracts, Report Series, Publication of the Department of Physical Sciences, University of Turku, 1994.

33. *Die Welt als Schnittstelle* ("The world as interface," with R. Rössler and P. Weibel, 1993)
In *Vom Chaos zur Endophysik — Wissenschaftler im Gespräch* (F. Rötzer, ed.), pp. 369–381 (Munich: Boer, 1994).

34. Intra-observer Chaos — Hidden Root of Quantum Mechanics? (1993)
Chaos, Solitons & Fractals **4**, 415–421 (1994).

35. Interfaciology (1993)
In *Hermeneutics and Science*, Proc. Int. Conf. Veszprém, 6–9 pp. 1993 (M. Fehér, O. Kiss and L. Ropolyi, eds.) (Dordrecht: Kluwer, 1997), in press.

36. Jumping Identities of Particles (1993)
Symmetry: Culture & Science **7**, 307–319 (1996).

37. *Das Jetzt als Interface* ("The Now as an Interface," abstract, 1994)
In *Nonlocated Online — Territories, Incorporation and the Matrix* (Zeitschrift Medienkunst), p. 5 (Vienna: Passagen Verlag, 1995).

38. Deterministic Continuous Molecular-Dynamics-Simulation of a Chemical Oscillator (with H. H. Diebner, 1994)
Z. Naturforsch. **50a**, 1139–1140 (1995).

39. Ultraperspective and Endophysics (1994)
BioSystems **38**, 211–219 (1996).

40. Space-Discretized Verlet Algorithm from a Variational Principle (with W. Nadler and H. H. Diebner, 1994)
Z. Naturforsch. **52a**, 585–587 (1997).

41. Subjective Objectivity (1995)
Found. Sci., submitted.

42. Interface Physics — Is the Observer Diameter Encoded in the Fundamental Constants c and h? (with R. Rössler and P. Weibel, 1995)
J. Consc. Studies, submitted.

43. *Ich bin, doch ich habe mich nicht* ("I Am but I Do Not Possess Myself," 1995)
In *Ernst Bloch Memorial Volume* (J. R. Bloch, ed.), pp. 262–275 (Frankfurt: Suhrkamp, 1997).

44. Relative-State Theory — Four New Aspects (1995)
    Chaos, Solitons and Fractals **7**, 845–852 (1996).

45. *Chaos — eine kopernikanische Wende?* ("Chaos — A Copernican Turn-about?", 1995)
    In *Komplexität und Sellstorganisation* (H. Krapp and T Wagenbauer, eds.), pp. 77–97 (Munich: Wilhelm Fink Verlag, 1997).

46. *Das Flammenschwert — oder Wie hermetisch ist die Schnittstelle des Mikrokonstruktivismus?* ("The Flaming Sword — or, How Hermetic Is the Interface of Microconstructivism?", 1995)
    Berne: Benteli Verlag, 1996.

47. Time Reversibility and the Logical Structure of the Universe (with H. H. Diebner and T. Kirner, 1996)
    Z. Naturforsch. **51a**, 960–962 (1996).

48. The Mystery of the Interface (in Japanese, 1996)
    Review of Today's Thinking (Japan) **24-11**, Sep. 335–346 (1996).

49. *Descartes 400 Jahre* ("Descartes 400 years," with F. Rötzer, 1996)
    In *Telepolis Magazine* (F. Rötzer, ed.) (1996); see: http://www.heise.de/tp/deutsch/inhalt

50. Rite of Passage: Only by Stepping out of Physics Is It Possible to Enter It (in German, 1996)
    In *Übergangsbogen und Überhöhungsrampe* (B. Ecker and B. Sefkow, eds.), pp. 130–133 (Hamburg: Material-Verlag, 1996).

51. The Relativist Stance (with K. Matsuno, 1996)
    BioSystems, in press.

52. Micro Time Reversal — Is Reality Observer-State Relative? (with H. H. Diebner and A. Kittel, 1996)
    Informatik-Forum (Vienna) **11**, 1, 4–6 (1997).

53. Micro Relativity (with H. H. Diebner and W. Pabst, 1996)
    Z. Naturforsch. **52a**, 593–599 (1997).

54. Cession — Twin of Action (in Spanish, 1996)
    In *Arte en la Era Electrónica* (C. Gianetti, ed.), p. 124 (Barcelona: Goethe-Institut Barcelona, 1997).

55. Vertical and Horizontal Exteriority — The Blue Card in Lampsacus (in German, 1996)
    In *Inteface 3* (K. P. Dencker, ed.), pp. 303–313 (Hamburg: Verlag Hans Bredow Institut, 1997).

56. On Exo and Endo Symmetry (in German, with P. Weibel, 1996)
    In: P. Weibel, *Jenseits von Kunst* ("Beyond Art"), p. 225 (Vienna: Pas-

sagen-Verlag, 1997).

57. Can a Program Force the Programmer to Reply? (1997)
    In *Art @ Science* (C. Sommerer, ed.) (Berlin: Springer-Verlag, 1997).
58. Anaxagoras' Unmixing of Chaos Through the Spirit of Ethics (in German, 1997)
    In *Lab 1996/97* (H. U. Reck, S. Zielinski, N. Röller and W. Ernst, eds.), pp. 277–292 (Cologne: Verlag der Buchhandlung Walther König, 1997).
59. Endophysics and Satori (in Japanese, with K. Matsuno, 1997)
    Contemporary Philosophy (Japan), December issue, 1997.
60. A Deterministic Entropy Based on the Instantaneous Phase Space Volume (with H. H. Diebner, 1997)
    Z. Naturforsch. a, submitted.

Index